Macmillan Computer Science Series

A. Abdellatif, J. Le Bihan and M. Limame, *Oracle – A User's Guide*

Ian O. Angell, *High-resolution Computer Graphics Using C*

Ian O. Angell and Gareth Griffith, *High-resolution Computer Graphics Using FORTRAN 77*

Ian O. Angell and Gareth Griffith, *High-resolution Computer Graphics Using Pascal*

C. Bamford and P. Curran, *Data Structures, Files and Databases, second edition*

P. Beynon-Davies, *Database Systems*

P. Beynon-Davies, *Information Systems Development, second edition*

Linda E.M. Brackenbury, *Design of VLSI Systems – A Practical Introduction*

Alan Bradley, *Peripherals for Computer Systems*

P.C. Capon and P.J. Jinks, *Compiler Engineering Using Pascal*

B.S. Chalk, *Computer Organisation and Architecture*

E. Davalo and P. Naïm, *Neural Networks*

S.M. Deen, *Principles and Practice of Database Systems*

Joyce Duncan, Lesley Rackley and Alexandria Walker, *SSADM in Practice*

D. England et al., *A Sun User's Guide, second edition*

Jean Ettinger, *Programming in C++*

J.S. Florentin, *Microprogrammed Systems Design*

Michel Gauthier, *Ada – A Professional Course*

M.G. Hartley, M. Healey and P.G. Depledge, *Mini and Microcomputer Systems*

M.J. King and J.P. Pardoe, *Program Design Using JSP – A Practical Introduction, second edition*

Bernard Leguy, *Ada – A Programmer's Introduction*

M. Léonard, *Database Design Theory*

David Lightfoot, *Formal Specification Using Z*

A.M. Lister and R.D. Eager, *Fundamentals of Operating Systems, fifth edition*

Elizabeth Lynch, *Understanding SQL*

G.P. McKeown and V.J. Rayward-Smith, *Mathematical Foundations for Computing*

Tom Manns and Michael Coleman, *Software Quality Assurance, second edition*

B.A.E. Meekings, T.P. Kudrycki and M.D. Soren, *A Book on C, third edition*

R.J. Mitchell, *C++ Object-Oriented Programming*

R.J. Mitchell, *Microcomputer Systems Using the STE Bus*

R.J. Mitchell, *Modula-2 Applied*

J.P. Pardoe and M.J. King, *Object-Oriented Programming Using C++ – An Introduction*

Pham Thu Quang and C. Chartier-Kastler, *MERISE in Practice*

Ian Pratt, *Artificial Intelligence*

F.D. Rolland, *Programming with VDM*

S. Skidmore, *Introducing Systems Analysis, second edition*

Continued overleaf

Macmillan Computer Science Series
Series Standing Order
ISBN 0–333–71697–3 hardcover
ISBN 0–333–69339–6 paperback
(outside North America only)

You can receive future titles in this series as they are published by placing a standing order. Please
contact your bookseller or, in case of difficulty, write to us at the address below with your name
and
address, the title of the series and the ISBN quoted above.

Customer Services Department, Macmillan Distribution Ltd
Houndmills, Basingstoke, Hampshire RG21 6XS, England

Introducing Systems Analysis

Steve Skidmore

Second Edition

First published 1994 by Blackwells Publishers
108 Cowley Road
Oxford OX4 1JF

Published 1997 by
MACMILLAN PRESS LTD
Houndmills, Basingstoke, Hampshire RG21 6XS
and London
Companies and representatives
throughout the world

ISBN 0-333-72141-1

A catalogue record for this book is available
from the British Library.

This book is printed on paper suitable
for recycling and made from fully
managed and sustained forest sources.

10   9   8   7   6   5   4   3
05 04 03 02 01 00 99 98

Printed and bound in Great Britain by
Antony Rowe Ltd, Chippenham, Wiltshire

Contents

Preface

This book is the first of two texts about the development of computer systems. It presents a series of models and skills that should help the definition and delivery of appropriate, effective, maintainable and flexible information systems.

This first book is primarily concerned with what is often termed physical and logical analysis.

The book is primarily aimed at

- Undergraduate and Higher National Diploma and Certificate students undertaking a module in Systems Analysis and Design.

- Students undertaking professional examinations.

- Trainee Analysts studying for Professional qualifications or following Professional Development schemes.

- Practising Systems Analysts.

The text concentrates upon the activities of Systems Analysis and Design and the skills and attitudes required to undertake them. It is not concerned with the organisation of the tasks. The terms analysis, design and development are used fairly interchangeably to describe the whole system activity. In fact the arbitrary distinction between analysis and design is probably counter-productive.

The term user is employed in the sense of embracing the clients, operators and victims of the computer system. The term is too standard for its replacement in a textbook.

When I first contributed to this book I was an academic lecturer. For the last five years I have returned to being a full-time practitioner. The upside of this is that the models, methods and ideas introduced in this book have been successfully used in practice. The downside is that the gestation period has been much longer, for which I must apologise to NCC-Blackwell. The academic exercises have now all been moved to Appendix A which is based around a practical case study. Suggested answers to the case study questions are available to bona-fide lecturers directly from The National Computing Centre.

1 Context

1.1 INTRODUCTION

In 1979 Tom de Marco wrote a seminal book that was largely responsible for introducing structured techniques to the information systems (I.S.) world. In it he characterised the non-structured specification, the document intended to communicate the proposed systems to the user, as a textual Victorian novel, unloved, unread but not understood. Despite the fact that this specification essentially formed a contract between the user and the developer he found that the user was unable to comprehend its content and hence would sign off the requirement with a shrug of his shoulders and an ill-founded hope that the computing department knew what it was doing.

Unfortunately by the late seventies considerable evidence was amassing to suggest that data processing departments were not confident in what they were doing. Budgets and timescales were overrun, systems were delivered that lacked important functionality or were unwieldy or unreliable to use. Evidence also emerged about the considerable amount of time these departments spent modifying or 'maintaining' systems that had already been delivered to users. Some of these changes were due to legitimate alterations in user requirements. However it was also clear that a significant proportion of changes were due to misunderstandings in the original specification and these had to be fixed before the system was acceptable to users.

The proponents of structured techniques recognised that the methods and models used to describe proposed systems to users were clearly faulty. Tom de Marco suggested that computer systems development badly needed a model, or set of models, which would allow the user and designer to confidently agree the functionality and scope of the system before the lengthy process of program development began.

It was felt that these models should fulfil the following criteria.

— *Graphical*. Conventional specifications were overwhelmingly textual. Diagrams were normally limited to describing a proposed technical solution, showing the location of workstations and the links between them. If a picture is worth a thousand words the developer unfailingly chose the thousand words.

— *Logical*. Physical descriptions of current business practices and the proposed computer hardware configuration occur at the start and end of the development cycle respectively. For the most part the system designer should focus on the logical models that describe the business systems stripped of their current and proposed physical implementation. In this way the project considers what the business needs to do – not how it does it now or how it will do it in the future.

— *Business oriented.* Specifications should be presented in terms of business rather than computer systems. It is impossible, and in fact undesirable and unrealistic, to expect users to sign off technical specifications which describe the business configuration in terms of the hardware solution and its associated technical implementation.

The 1980s saw the gradual, and sometimes grudging, acceptance of structured techniques as an analysis and specification method for users and developers alike. Formal methodologies, whether they are published (such as SSADM, LSDM and Information Engineering) or developed in-house are gradually displacing 'craft' and 'experience' based notions. This adoption of formal techniques has been assisted by the emergence of powerful CASE tools to describe the essential logical models.

Four other trends have also affected the practice and teaching of systems development in the last decade.

— The increased power of software and the continued emergence of new Fourth Generation Languages has meant that more attractive and flexible solutions can be quickly developed and delivered to users.

— The user community has become more comfortable with Information Technology. The growth of home computing and business PCs has led to more understanding (and hence less fear) of the technical issues of system delivery.

— Economic recession has led to an even greater need for businesses to secure value for money from systems investment. System developers need to show that they are economically delivering effective systems.

— A staggering decrease in hardware (and to a lesser extent) software prices has meant that it is economically feasible to computerise many areas of the enterprise. Reduced reliance on proprietary Operating Systems has meant that IS has greater control over its own destiny.

This book reflects these changes in method and environment. It uses formal techniques to describe and specify requirements. These techniques are taken from different methodologies (although SSADM notation is used as a default) and some have been around a long time; such as decision tables and System Flowcharts. Others, such as Object Oriented development methods, are emerging as part of the continued development of formal specifications to reflect new possibilities, requirements and constraints.

The approach here is to provide general analysis training and to select techniques from certain methodologies when they suit the purpose at hand. The overall aim is to give the practitioner a set of tools which can be selected to suit different circumstances. The principle of this tool-kit approach has been elaborated elsewhere (Benyon, 1987) and will be reviewed at the end of the companion text *Introducing Systems Design*.

However, it must be stressed that these techniques are used because they work and hence are worth teaching and examining. I am still aware of developers who feel that they can somehow construct systems using common-sense, experience and general business knowledge. I have worked with these developers. Their projects are usually unsuccessful.

1.2   THE SYSTEMS DEVELOPMENT LIFE CYCLE

The Structured System Development Life Cycle model is a convenient place to start an examination of what an analyst actually does. This model (see Figure 1.1 for our version) suggests a number of stages:

— *Strategic Study.* This is concerned with defining possible IS contributions to the objectives of the enterprise. A number of candidate applications will be identified which will then be subjected to detailed Feasibility Studies.

— *Feasibility Study.* Each candidate area is examined and outline business and technical solutions suggested. A comparison is undertaken in the context of economic, technical and operational issues, culminating in a Feasibility Report which recommends a possible solution and comments on whether detailed analysis should commence. Candidate areas are compared within the organisation to see which are worth investing in and from this detailed systems projects are initiated. Feasibility Studies usually use the models and techniques introduced in Chapters 4 – 7 of this book.

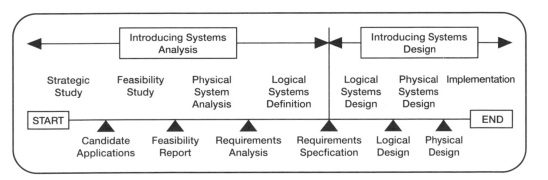

Figure 1.1   System Development Life Cycle – showing analysis stages, deliverables and the scope of the book

— *Physical Systems Analysis:* Once the scope of the study has been agreed, a detailed investigation is undertaken of the operations of the current system and the requirements of its successor. This culminates in a description of the current physical systems using the models described in Chapter 5 of this book and supported by documentation introduced in Chapter 4. The document produced at the end of this stage is often termed the Requirements Analysis.

— *Logical Systems Definition.* The required system is then developed using models from a data (Chapter 6), process (Chapter 7) and event (Chapter 8) perspective.

This is supported by a common Data Dictionary (Chapter 9). The document signed-off at the end of this stage is usually called the Requirements Specification.

— *Logical Systems Design.* This will primarily be concerned with logically defining the data structures (normalisation) and defining detailed processes (action diagrams, structure charts and state transition diagrams). This is covered in the complementary textbook *Introducing Systems Design.*

— *Physical Systems Design.* This stage develops physical inputs and outputs, files, programs, databases and controls. It is again covered in the complementary textbook *Introducing Systems Design.*

— *Implementation:* The testing of programs and systems and the development of supporting manuals and documentation. Implementation also includes the phasing in of the new system using an appropriate management method and the organisation and conduct of training courses.

— *Maintenance:* The implementation of amendments and omissions. These are likely to increase as time passes and new requirements and possibilities emerge as the environment of the system changes and new hardware and software opportunities present themselves. This culminates in the consideration of a new system, a Feasibility Study is commissioned, and the development cycle begins again.

1.3   STRUCTURE OF THE BOOK

The book begins with an examination of the role of information systems in the context of the enterprise (Chapter 2). Every organisation faces at least two fundamental problems when considering computerisation.

- Identifying relevant applications.

- Giving appropriate priority to these relevant applications.

This can only be done successfully if the business contribution of competing applications is properly understood. Chapter 2 examines the setting of business objectives, the formulation of a strategic plan and the contribution of information systems to both planning and objective setting. It also recognises the implications of the work of Mintzberg and Checkland and summarises the latter's soft-systems approach.

Candidate applications for computerisation may now be subjected to a detailed Feasibility Study, and this is examined in Chapter 3. The costs of technical and operational feasibility are acknowledged and four ways of assessing economic feasibility are presented.

Chapter 4 looks at some of the inter-personal skills needed by developers to successfully analyse and implement systems. There is a trend to present systems development as a technical skill; the construction of a set of increasingly complex models that will guarantee success. This is probably a reaction to past conventional training which tended to over-stress the personal characteristics of the analyst, placing too much emphasis on interviewing, conversational and presentation skills. There needs to be a balance between technical modelling and inter-personal skills with the latter seen as supporting the former.

An understanding of the current information systems that support the organisation gives a fundamental insight into the design requirements of any successor. Chapter 5 develops two models for describing the current systems – the Systems Flowchart and the Data Flow Diagram.

However, there is a problem of moving from an understanding of the present system to the design of a replacement. Relying on current procedures can undoubtedly lead to designs that fail to harness the power of the computer and lead to underperforming 'computerised manual systems' whose restricted scope reduces their chance of success.

This transition from Analysis to Design has been recognised by the growth of logical modelling – techniques that show the logical information system requirements stripped of their administrative arrangements and physical trappings. These models have had a significant impact on the way that computer systems development has been taught and practised. The following three chapters examine complementary logical models.

Chapter 6 shows how physical Data Flow Diagrams can be developed into their logical equivalents. The Data Flow Diagram is a central technique of most of the Structured Methodologies, although notation and nomenclature vary from vendor to vendor. It is an effective way of modelling the system from a process perspective.

Chapter 7 examines a technique that takes a more static view of the data. The Logical Data Model provides important insights into the data required and maintained by the organisation. This data driven view complements the process driven perspective of the Data Flow Diagram.

Chapter 8 introduces the event perspective, showing how the entities of the system are created, modified and deleted. The Entity Life History (ELH) is used to model this perspective.

Chapter 9 describes various supporting models of which the most important is the Data Dictionary. This provides a facility for the management of the data resource. It is emerging as not only an important documentor of systems, but also as a very active tool in the whole system development spectrum.

Chapter 10 provides a number of alternative perspectives.

- An extension of the Data Flow Diagram to allow more sensitive and accurate modelling of real-time systems.

- An alternative notation for data modelling. This is the Chen notation used in many data modelling texts.

- Prototyping

- Object Oriented Analysis

Finally, Chapter 11 looks forward to the companion systems design book. Figure 1.2 describes the succession of models used in this text.

1.4 SUMMARY

This chapter has introduced the context of contemporary analysis and the framework of this book. It must also be recognised that systems development is continually changing. A nod towards these changes is presented in Chapter 10 where object oriented analysis is briefly considered. However, the emphasis of this book is on established structured techniques and models.

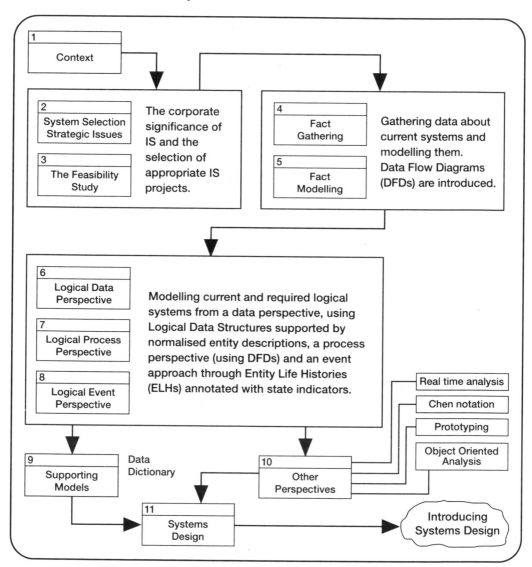

Figure 1.2   A Road Map of the Text.

2 System Selection: Strategic Issues

2.1 INTRODUCTION

This chapter looks at the early stages of a project, examining some of the issues that will be encountered in project initiation and selection.

It suggests that project selection should reflect the business priorities of the enterprise and that information systems development should take place within the context of the business strategic plan.

2.2 PROBLEM IDENTIFICATION AND SELECTION

Every organisation faces two fundamental problems when it considers the role of computers.

- It has to identify opportunities for computerisation.

- It needs to give these opportunities appropriate priority.

It is important not to jump to conclusions when establishing areas of the enterprise that might benefit from computerisation. Payroll, for example, is a common and relatively simple application, but there is no reason why this should necessarily be computerised in every organisation. The fact that something can be computerised does not mean that it should be. Computer systems that aid the organisation's aims and prosperity should receive priority. A simple, slightly fictionalised, example illustrates this.

LDF Engineering was a small engineering firm based in the Midlands. It appointed a keen and aggressive Accountant who set about the task of acquiring the firm's first computer to support his planned computerisation of the accounting ledgers and payroll. The hardware and software selection was performed meticulously and an appropriate system was purchased and installed. Just over one year later the firm went into voluntary liquidation. At the post mortem it was felt that over-stocking of certain product lines and poor production planning and control had been major contributors to the company's decline. Yet these are applications suited to computerisation! In retrospect, the firm appeared to have successfully computerised the wrong system. If better Stock and Production controls had been applied through using a computer system they may have resulted in something that the accounts application could not offer – the organisation's survival.

The selection of relevant applications is undoubtedly much easier in organisations where information systems planning is seen as an integral part of the enterprise's overall planning effort. In such instances the computer systems are recognised as a valuable resource within the organisation's total activity not as an appendix to it. If enterprises fail to have any overall business planning, then it is difficult to formulate a coherent information system strategy. This corporate view of computing was central to the Effective Computer approach developed by Grindley and Humble (Grindley, 1973) and based upon the principles of Management by Objectives. Their book was written in response to observations of many projects where the computer failed to satisfy the expectations of managers, a problem that still exists two decades later.

They observed that:

"The only valid objective for computers is to assist in achieving defined business improvements which would be impossible or uneconomic without the computer".

Thus the computer is viewed as a resource to help the doers and makers of an organisation achieve their profit or service objectives.

2.3   STRATEGIC PLANNING –
AN INFORMATION SYSTEMS PERSPECTIVE

If appropriate system development is easier in an organisation with a coherent plan, then it is important that every analyst should have some insight into the Strategic Planning activity.

"Planning is the design of a desired future and of effective ways of bringing it about" (Ackoff,1970).

Strategy is concerned with the general direction and long-term policies of the enterprise. The organisation defines where it wishes to be in terms of its objectives and where it is currently, through a corporate appraisal. Moving from the current to the desired state is the task of strategic planning.

2.3.1   The Importance of Objectives

Strategic Planning texts use different definitions of objectives, goals and targets. This book uses an adaption of Hussey's four-stage hierarchy of objectives.

A primary economic objective.

This is an overall target for the company established in advance of how that target is to be achieved. Examples might include:

— To treble after tax profits from £250,000 to £750,000 by the end of the fifth year of the plan.

— To provide a £50,000 income package for each of the Partners by the April of the second year of the Plan.

Many writers argue that most primary objectives will be established in terms of profits (revenues in excess of costs) or profitability (return on investment). This is because even if the company has other objectives, it must ultimately make a profit or return on investment, or sooner or later it will cease to exist.

Secondary objectives that describe the intended corporate identity.

These give a picture of the company's aims in terms of the nature and scope of the business, its trading and staff relations and intended developments, expansions and locations. In this respect it is a statement of how the Chief Executive, Chairman or Directors perceive the future of the company. The narrative objectives are a method of trying to communicate and share that perception with the rest of the company.

These social and non-economic objectives reflect the needs of the firm's stakeholders, such as employees and shareholders, and they serve to influence and constrain management's pursuit of the primary economic objective.

Goals or intermediate objectives

Goals are milestones on the way to the overall economic target. A network of compatible goals leads to the achievement of the primary objective.

Goals might be established for.....

— The capture of a certain market-share.

— An absolute value of sales.

— A minimum figure for customer complaints.

— A cost reduction target.

— A minimum time for answering an emergency call.

— A date by which a product launch must take place.

Finally, Standards of Performance are given to particular individuals.

This will often be achieved by splitting up the corporate goals. Thus, for example, an overall sales figure may be divided into territories, a profit target, into departments, etc. A set of Performance Standards will be compatible with a goal which, in turn, will complement the other goals contributing to the overall primary objective of the company. This gives a direct link between the expected performance of the individual and the corporate task.

2.3.2   Contribution of Information Systems to objectives

The second stage of Grindley and Humble's Effective Computer approach established corporate objectives and identified how computer systems might contribute to achieving them. A series of examples are given, two of which are repeated below.

Corporation   Aircraft Spares
Key Area   Market-share

Objective   To increase sales volume by 40 percent by...
Contribution To install an on-line terminal in customer's premises showing our stock and delivery position and price.

Corporation Carton manufacturer
Key Area   Productivity
Objective   To reduce waste on box manufacture by 8 percent..
Contribution To implement an order analysis system for determining optimum width and run size for corrugated paper.

The strength of this business approach to information system development is that it places the computer system into the perspective of corporate objectives, and this is appropriate because these are the things that actually matter. The computer is seen as an aid to achieving some new business goal, one that would be difficult, costly or impossible to achieve without computerisation.

"Note that this approach differs from those which simply attempt to perform some existing system better (cheaper, faster, more accurately, etc.)" (Grindley 1973)

The potential contribution of information systems (IS) to the doers and makers of an organisation can be reviewed within a framework of how enterprises develop their competive strategy.

There are essentially four competitive strategies (Ansoff, 1987) and these are examined below in the context of information systems.

1   *Reduction of the cost of a product or service.*

Micro-economic theory hypothesises that market success largely depends upon the price of a product or service. It suggests that the enterprise with the lowest price will become dominant in the marketplace and this will lead to discounts and economies of scale which, in turn, will reinforce its position as the producer with the lowest costs.

This competitive strategy is referred to as the market-share or market-position strategy and the result is a strategy that minimises costs to allow the product to be sold at a price equal to or less than the competition.

Computerisation may contribute to this strategy by reducing;

(i) the direct unit cost of a service or product.
(ii) the overhead costs associated with the production of that product or service.

Information systems (the focus of this book) are usually associated with the latter set of cost reductions. A typical example is provided by Scholes and Klemm (1987) who cite an engineering company manufacturing a range of pneumatic tools and pumps. Production control systems implemented in 1982 at a cost of £200,000 contributed productivity savings of 30 percent and a stock reduction of 20 percent in that year. The total saving was estimated at £500,000.

"The company had become leaner and fitter, not through cutting back the workforce but by cutting down surplus stock."

2 *Increase revenues by offering a competitive advantage*

A second competitive strategy is to provide a product/service differentiation which permits the company's product or service to be successfully differentiated from its competitors.

This differentiation is often achieved through service or product quality and usually provides an organisation with a competitive edge.

For example, three companies compete for timber handling, storage and distribution in a busy northern port. There is no cost differentiation between the companies because a standardised rating system is used. The manager of one of the companies recognised that a stock importing, storage and distribution system would help him meet his primary objective of increasing market-share to 50 percent of the quay's timber business. The system would allow him to provide up-to-date information about the customer's current at-sea, stock and delivery position and this would give him a competitive edge over his rivals. He reasoned that he would become their preferred supplier if he could provide this information.

In many markets customers expect to be provided with accurate and up-to-date information. This may not be essential for the running of the business but a failure to offer such a service is likely to reduce the company's competitive position.

After the successful installation of the timber handling system and the achievement of the required market-share, the company turned its attention to increasing market penetration with particular customers. They recognised that certain customers placed 20–30 percent of their trade with competing companies despite their delight with the information provided by the new system. Consequently the second stage of the system provided customers with a direct link into the system so that customers could enquire and instruct without contacting the timber company. It was felt that customers with a direct input into the system would be unlikely to continue using the ponderous manual procedures of the competitor companies.

This approach effectively ties customers into the provider's information systems and hence achieves a competitive advantage by making it difficult for customers to switch suppliers.

3 *Provide a market or image differential*

Experience has shown that purchasers are often affected by the image of the products or services offered. Enterprises began to recognise that customer needs could be shaped through marketing and promotion aimed at projecting an image to appeal to a particular market niche. This strategy of market differentiation "defines the way in which the firm will differentiate itself from competition in the eyes of the customer" (Ansoff 1987).

Systems may be developed to help create and reinforce this image through improving product presentation and quality.

4 *To provide organisational growth*

The competitive strategies examined so far are primarily concerned with determining how the firm goes about achieving the objectives it has set itself in its selected business areas. A related strategic issue concerns the selection of those areas in which it wishes to undertake business in the first place. This selection is often termed portfolio analysis and it is concerned with an analysis of the product and market opportunities available to an enterprise beyond the scope of its current activities. Portfolio analysis will help decide whether the firm should diversify or expand in some way in order to achieve its primary objective.

In undertaking portfolio analysis IS can help in at least the following ways;

(i) IS managers can help identify new products and markets which can then be subjected to portfolio analysis.

It is clear that this has happened effectively in a number of companies where systems developed for internal use have been extended into packages and offered, often through an associated company, into the wider marketplace. Many engineering and manufacturing organisations have spawned successful subsidiary software companies.

(ii) IS can offer operational synergy to organisations, hence increasing the attraction of entering a particular marketplace. Synergy is often defined as 2 + 2 = 5; the sum of the whole is greater than the sum of the two halves.

A company specialising in the import and sale of timber, subcontracted its storage and distribution because of the high administrative overhead of its core business. However, the successful computerisation of the import side meant that business diversification became possible and various marketplaces were considered. Discussions with software suppliers showed that the system developed for the core import and sales business could be extended relatively easily and economically to cover the storage and distribution sides. As a result the company terminated its sub-contracts and began to re-position itself as a one-stop supplier of timber from mill through to builder's yard.

(iii) IS can develop models to support portfolio analysis itself.

The Sunday Times 'IT for Business Excellence' awards also illustrate further examples of IT contributions:

— Severn Trent Water's investment in IT systems has dramatically improved responses to customer's enquiries about bills. The average response time has been reduced from 16.5 days in April 1990 to 5.4 days in December 1992.

— IT has enabled Nationwide's new customer service team to handle enquiries relating to 95 percent of all mortgage and insurance transactions without specialist support, significantly speeding up reponse times.

— Greenup and Thompson is a Sheffield packaging company. Following the installation of a management information system coupled with computer-aided design and manufacturing facilities, it reduced production lead times from 6-8 weeks to 3-5 weeks. Samples of new carton designs can now be produced in 30 minutes compared with 90 minutes previously.

— From a standing start in January 1993, National Westminster Life Assurance's field salesforce submitted more than 10,000 insurance and pensions proposals in the first month of operation – all through electronic point of sale systems. The company reports that IT significantly improves the turnaround time for proposal acceptance and issue, enabling it to

"set service standards competitive with the best in the industry".

all examples – Sunday Times, 14 March 1993

Thus information system strategies have been identified in a 'top-down' way, moving from an overall corporate target to individual goals which might be achieved with the aid of a computer system. This should ensure a unity of purpose impossible with piecemeal and unfocused system development. Three important points about objectives should be reinforced.

— It is important that the objectives are quantified. This forces realistic target setting and agreed measures of success. An objective to 'reduce inventory holding' is successfully achieved if only one less item is carried! However, an objective to reduce inventory by 5 percent by January 1997 makes the criteria of success much more specific. Time constraints are also important for objectives, so that they not only indicate what should have resulted but by when it should have taken place.

"A main difference between companies which practise effective formal planning and those which have a more traditional approach to management is that the planning companies are not satisfied with words alone. Much more meaningful is a specific quantitative statement of what profit is required."

(Hussey 1982)

— Objectives can be set for non-profit making organisations. These might include:

• All emergency calls will be answered in 2 hours.

• No patient will have to wait more than 30 days for throat surgery.

• 80 percent of our students will pass the Level 2 examinations.

— Objectives are set for a particular **business** goal, not for the computer's contribution. Any attempt to partition objectives, or indeed success, between the computer and the non-computer part of the corporate effort is doomed to failure. The important matter is to achieve objectives not to squabble about who, or what, achieved them.

2.3.3   Corporate Appraisal

A corporate appraisal of the enterprise's current strengths and weaknesses is an important first step in the planning process. It sets a baseline for planning as well as providing an opportunity to critically examine established beliefs, operations and practices. It may also directly identify means for profit improvements.

Hussey (Hussey, 1982) identifies ten major aspects of a corporate appraisal and some of these have important IS implications.

1   *Trends of results.* The company's historical pattern of performance. This will cover trends in profit, sales, capital employed, as well as all the various financial ratios that can be applied to measure efficiency and performance. This will give an overall impression for whether the company is improving or worsening its position. Declining or troubled companies may impose tighter constraints on IS development and expect quicker returns. Alternatively, and perhaps more dangerous, they may not invest enough in system development (preferring to attack what they consider to be the 'real' – i.e. immediate – problems), and so miss technological opportunities.

2   *Sources of Profits.* This will consist of a detailed examination of profitability and prospects. Such a study aims to identify the actual 'breadwinners' of an organisation and this may lead to product variety reduction, elimination of certain products, etc. If this study is not undertaken then IS development time and resources may be wasted upon products and problems which are not important in the production of profit or the delivery of services.

3   *Risk.* An examination of the risk associated with the company's sources of profits. This is of particular significance to enterprises that are very dependent upon one product or a few customers or suppliers for trade. Effective information systems are **critical** in high risk markets or circumstances, so that the very damaging effect of detrimental changes can be quickly identified and corrective action attempted.

4   *Manufacturing activity.* This considers the product processes and identifies possibilities of new production opportunities. The traditional commercial basis of much IS development has undoubtably led to a lack of consideration of production and design opportunities.

5   *Rationalisation of resources.* A study of whether plants, depots and buildings cannot be amalgamated in some way. This can be applied to computer facilities. There may be much to gain by using such resources in a more cost-effective manner.

6   *Organisation and Management.* The organisational structure of the company and its strengths and weaknesses. This will include problems that might appear in the future such as retirements and difficulties of succession. Managerial capabilities have to be assessed and their strategic implication understood. As Hussey comments...

"There is very little point in deciding to launch a range of new products if it is known that the marketing manager is incapable of making a success of the venture."   (Hussey, 1982)

The importance of corporate structures to the IS developer is considered in the next two chapters. However, Hussey's point is valid at the strategic level of project selection. An inventory management system does not, **in itself**, lead to a reduction in stock levels, it merely facilitates it. An incompetent Stores Manager might only make poor decisions faster and at greater cost if supported by a newly installed information system.

The general 'climate' of the company is also important. Computer systems usually bring significant changes and the analyst must assess how those changes are likely to be viewed by employees and, furthermore, what mechanisms for implementing change already exist in the company.

7   *Financial Resources.* An assessment of the company's capital and liquid resources and current and projected cash flow. A company with a very high debt looks at the future very differently to one with a 'cash mountain'. Economic feasibility (discussed in the next chapter) is critically affected by this issue. The financial position determines how systems are funded, when they will be paid for and, most important, how much is available for their development. A clear understanding of the financial position is required if projects are to be sensibly assessed and selected.

8   *Corporate Capability.* A review of the strong and weak points of the corporate operation: "what we are good and bad at doing and why". This will again aid project selection.

9   *Systems.* An appraisal of the strengths and weaknesses of the operational systems which support the company. This will not only look at current efficiency, but also the effect of likely changes...

- "Can the order processing department cope with the projected increase in orders for this product?"

- "What are the picking routines in the warehouse?"

- "How are marketing promotions decided?"

This assessment is clearly of vital strategic importance to the IS section. It may encompass systems already under its jurisdiction and subject their performance to searching corporate review. Furthermore, it is likely to identify possibilities for further computerisation and incorporate these in the business objectives of the resultant Strategic Plan.

10 *Use of Resources.* Identifying how the various resources – people, money, plant, etc., are brought together to produce the company's products. This will include a consideration of whether the adopted pattern is the most effective one. Are the right people employed in the right place? Is the product emphasis correct?, etc. One of these resources is information systems and such an appraisal may lead to radical changes in organisation, for example, from centralised to de-centralised processing.

Thus the corporate assessment is a wide ranging examination of the company's position. Information systems development will be affected by all aspects of this appraisal and fundamental insights and 'ground rules' will be established. It provides a statement of the current position of the enterprise.

The Strategic Plan of the company is then formulated to move the organisation from its current position (defined in the corporate appraisal) to its desired situation (described in terms of its objectives).

The formulation of corporate objectives and the plan developed to achieve these objectives will be sanctioned by the Board of the company or its equivalent. If

Information Systems staff are not included at this level then it is likely that IS staff will feel totally excluded from business policy. This usually leads to frustration and a belief that IS's contribution is both undervalued and misunderstood. As a result many staff begin to identify with technical and professional aims rather than business ones. This behaviour, in turn, reinforces the reasons for their exclusion from the management of the business.

Consequently, there are often demands for the appointment of an IS Director at Board level. This is a formal response to the problem of getting IS input into key business decisions.

2.3.4   Strategy crafting

Viewing information system development from the strategic business perspective is both useful and valid. However, it does provide a rather clinical and idealistic image of the current and future states of the organisation. Enterprises tend to consist of many inter-relationships, fired by complex webs of operations, conflict and deceit. The actual enterprise is much messier than the rather sanitised version that often emerges from the planning document.

Mintzberg (1989) suggests that the 'top-down' approach is largely unsupported by observation and that in fact potential strategies grow in all sorts of places – like "weeds in a garden". In fact these potential strategies take root in all places where employees are in touch with the actual situation of the company and have time and resources to consider improvements.

He suggests that large shifts in the strategic direction of the company result when the enterprises current strategic direction becomes out of sync with its environment, usually by threatening the achievement of the primary objective. At this time management can begin to germinate some of the seeds of strategies held in check under the previous regime. This is generally a better alternative than having to develop new strategies from scratch or to import them from elsewhere. In this way the organisation turns emerging patterns from its own company into a new orientation.

If an organisation does determine strategy in this way then this has important implications for IS staff. If they are not contributing strategic visions to the enterprise then it seems likely that the potential contribution of IS will not be recognised. Indeed a common criticism of IS staff in general is that they have more allegiance to their profession than to their company. The result is their exclusion from potential strategic input and hence opportunities are missed. This approach also puts the internal IS department at risk because it becomes identified as a cost centre rather than a possible source of growth and profit.

The top-down approach also represents a managerial perspective of the organisation. This is clearly an important one, but it is not the only valid viewpoint. Not all employees concerned with information system change will share the management's perspective of the company and the analyst must be aware of this. Furthermore, management's view of reality may differ considerably from what is actually taking place. Descriptions of the same system given by managerial and operational staff may differ alarmingly in both detail and perspective.

Finally, many companies do not have a Strategic Plan, and of those that do, many do not consider information system development in its organisational perspective. In such instances there is no obvious starting place for project selection and priority.

Consequently we need to employ a framework that might concede the preceding points as well as giving an insight into Strategic Planning initiatives. The Soft Systems methodology of Checkland (Checkland, 1981) is a candidate framework.

2.4   SOFT SYSTEMS METHODOLOGY

The Soft Systems framework is primarily concerned with tackling the ill-structured problems of the real world and suggesting solutions that may, or may not, include computers. It attempts to tackle the fuzzy and complicated problems of organisations where there often appears to be insurmountable problems in defining the problem, let alone solving it. The framework was developed and presented by Peter Checkland (Checkland, 1981) and variations appear in Wilson (Wilson, 1984) and Wood-Harper, Antill and Avison (Wood-Harper, 1985). The overall model is given in Figure 2.1.

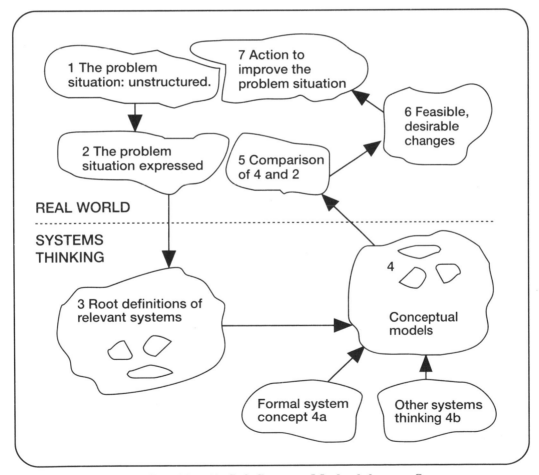

Figure 2.1   Checkland's Soft Systems Methodology: a Summary

The first two stages are concerned with examining a particular problem situation and expressing it without imposing a solution. The problem area needs to be explored as widely as possible building up the richest possible picture of the situation being studied.

At the beginning of a systems project the people involved are likely to have only a fuzzy idea of what they want to achieve. Even if the proposed objectives can be sorted out, they still need to be formulated in such a way that they can be explained to analysts and potential suppliers. A useful technique for modelling the overall system under consideration is the 'Rich Picture'. It attempts to show what the organisation is about.

The Rich Picture is useful to the analyst because it provides an overall view of the current problem area as well as giving greater understanding of it. Drawing a Rich Picture should:

— Focus attention on the important issues. It is common for analysts to become too concerned with detail too early on in a system study. Make no mistake, attention to detail is extremely important in information system design, and there are numerous techniques to support and encourage this level of precision. However, this may lead to the analyst 'not seeing the wood for the trees', or indeed, failing to recognise that his perceived wood is actually a small part of a large forest! The use of a Rich Picture, constrained to one side of A3 or A4 paper, demands concentration on major flows and structure, stripped of their detail. In this respect it mimics the understanding of the analyst. Users and operators are inclined to describe their tasks in detail and from this the analyst must try to glean the overall structure and process of what is taking place. Many participants in the system are, by their very proximity to the task, unable to perceive the whole activity in which they are involved and are consequently more able to describe what they are doing rather than why they are doing it.

— It will help participants to visualise and discuss the role they have in the organisation. The analyst has a model of the enterprise which may be incorrect, partial or biased. Inconsistencies and errors in this model are easier to sort out if the model is itself described in some pictorial form. Hence the Rich Picture can act as an important method of communication.

— It can be used to define the parts of the organisation which will be covered by the information system. The definition of a system boundary is an important task. The analyst and the participants need to be clear about what the proposed system will cover, and, perhaps more important, what it will not.

— It can be used as a medium for illustrating participants' worries, conflicts and responsibilities.

Rich pictures should be largely self-explanatory as they are designed to aid communication as well as help the analyst visualise the problem situation. It is a method of pictorially representing three important considerations in information systems design:

• Elements of structure in the problem area. This might include departmental boundaries, physical or geographical layout and product types and activities.

- Elements of process. What takes place in the system.

- Elements of relationships. The relationship between structure and process is the climate of the problem area. This will include conflicts, worries and mismatches between new processes and old structures.

Figure 2.2 shows an example Rich Picture.

Stage 3 begins at the end of problem expression and tackles the question: What are the systems which, from the analysis stage, seem relevant to the problem? This question has to be answered carefully and explicitly and the resultant Root Definition will also reflect a certain personal view of the organisation. A number of examples are given below to give a flavour of their formulation.

— A department to employ social workers and associated staff to build and maintain residential and other treatment facilities and to control and develop the use of these resources so that those social and physical needs of the deprived sections of the community which Government statute determines or allows, to the extent to which County Council, as guided by its professional advisers, decides is appropriate, are met within the annual capital and revenue constraints imposed by the Government and Council.

(Social Services Department)

— A system to celebrate a particular lifestyle using pop-music as an emblem of the sub-culture concerned.

(A Pop Festival)

— A system owned by the Manpower Services Commission and operated by the Paintmakers Association in collaboration with the Polytechnic of the South Bank's Distance Learning Unit, to provide courses to increase technical skills and knowledge for suitably qualified and interested parties, that will be of value to the industry, whilst meeting BTEC approval in a manner that is both efficient and financially viable.

(A Distance Learning Project at South Bank Polytechnic – Wood-Harper, 1983).

Root definitions can be developed for a number of perspectives of the same problem situation. In Checkland's approach the analyst is free to select any and many perspectives of the system and develop appropriate root definitions. The main characters of the GDL system perceive the problem situation in very different ways reflecting different Weltanschauung. In real life these views are often irreconcilable and hence the strategy of the company is affected by the relative bargaining strengths of the characters at different times.

Checkland distinguishes between primary-task and issue-based root definitions and acknowledges that companies/departments may have both.

For example, a study in a Management Service Department of a large manufacturing company had a primary-task definition to describe a system whose function was to provide intellectual services to support the management process

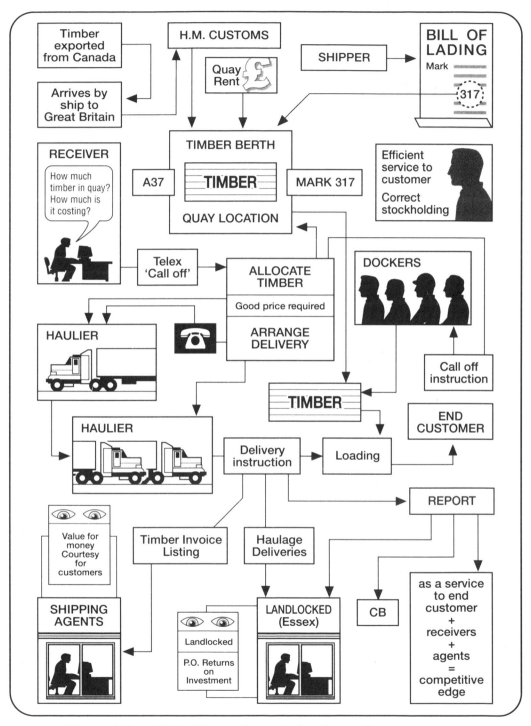

Figure 2.2   A Rich Picture for a timber importing, storage and distribution company.

in a turbulent business environment. However, a different root definition also acknowledged that any management services team would not be a direct wealth generator and hence would have to permanently justify its existence by demonstrating its usefulness. This latter demonstrates an issue-based root definition.

Root definitions are generally formed from the following elements

C	Customer	The client, beneficiary or victim of the system.
A	Actors	The agents who carry out, or cause to be carried out, the activities or transformations of the system.
T	Transformations	The core of the root definition. A process carried out by the system: assumed to include the direct object of the main activity verbs.
W	Weltanschauung	The (often unquestioned) outlook or taken-for-granted framework which makes this particular root definition a meaningful one.
O	Ownership	The owners of the system, control, concern or sponsorship.
E	Environment	Environmental impositions. Wider systems in which the transformations must operate.

Stage 4 makes and tests relevant conceptual models. This is concerned with what should be happening to support the requirements specified in the Root Definition: "the aim is to build an activity model of what must go on in the system" (Checkland, 1981).

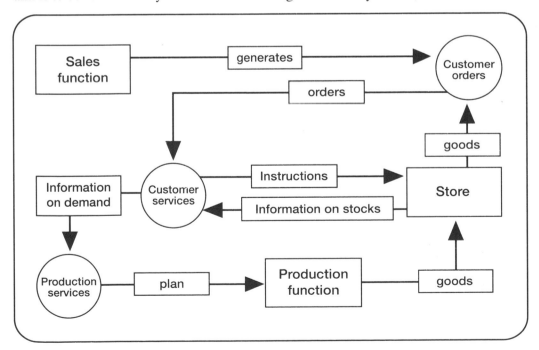

Figure 2.3 A Conceptual Model of an Order Generating-and-Processing System (from Checkland)

Many conceptual models are in fact quite similar for different types of organisation. The example given in Figure 2.3 would be relevant to many types and sizes of enterprises.

The next stage is to compare the conceptual model with reality. In effect this is examining the system that is required to support the Root Definition (Stage Four) with the one that actually exists (Stage Two). This will lead to the final two stages of implementing 'feasible and desirable' changes. Three kinds of changes are envisaged.

- Structural: Organisational groupings, reporting structures or functional responsibilities.

- Procedures: Processes of reporting and informing.

- Attitudes: Changes in influence, expectations and perception.

The strength of the Soft Systems approach is in its insistence upon a framework, rather than a prescription, for organisational activity. It tends to recognise the difficulty of getting to grips with a problem area and, in doing so, acknowledges a complexity which may be missing from the 'sanitised' corporate vision of the Strategic Plan.

The comparison of the actual situation against conceptual models is also very valuable. In Checkland's book he suggests that these models are essentially drawn from formal systems thinking or concepts. However, it is also now very clear that there are generalised IS models which essentially apply to many apparently different enterprises and situations. This point is taken up again in Chapter 10 when Object Oriented Analysis is considered.

Finally, traditional IS analysis and design results in recommended systems for information gathering, processing and outputs. Checkland's 'feasible and desirable' changes forces us to recognise that changes in procedure are only part of the story.

2.5 TRIGGERS OF CHANGE

Our discussion so far has concentrated upon strategic approaches to information systems development. The philosophy of Top-Down Strategic Planning is different to that of Checkland's Soft Systems approach. However, whatever their divergence they share the perspective of 'wholeness'. The perception that IS development must come from an understanding of the enterprise's whole activity and not from the piecemeal solution of problems. Indeed there are similar elements in the two approaches. Exploring the problem situation has much to do with corporate appraisal, whilst the conceptual model might be seen as part of the setting of objectives. It helps establish the desired state of the enterprise and hence will foster the formulation of objectives within the feasible and desirable changes identified.

The reasons why the organisation is considering change should be both well justified and understood. However, this trigger for change may not always be so well-defined. In such instances it is important that the analyst investigates the reasons for system development so that he can formulate the scope of the project and the climate in which he or she will be working. Three simple types of trigger may be identified.

Positive Trigger

"I spend a lot of my time compiling figures, producing and revising forecasts and then presenting reports and plans based on these figures. Surely the computer can help me in some way?"

In such instances the problems of the present system have been identified by those who run and use the system. The computer is seen as a possible way of getting system improvements. There is a positive commitment to computerisation – the system user is on the analyst's side.

Negative Trigger

"We have had problems with Stores ever since I became a Director. It's time we computerised it and got up with the times. That will sort them out down there!"

In many instances the decision to consider computerisation is taken by the system user, not those who actually operate the system. In such circumstances the perception of the present system's performance is unlikely to be shared by all those concerned with running the system. Indeed the operators of the system may see any investigation as a criticism of their present performance and the procedures that they have created.

In the example represented in the above quote, the company's managers had 'lost control' of Stores. They had been looking for an excuse to impose new controls for some time and computerisation presented that opportunity. All the members of the Stores section resisted the plans. They felt that management did not understand what happened in the Stores section and staff morale was low. They felt aggrieved that the systems that had been developed over many years were now deemed to be inadequate by people who, they believed, did not understand them.

In such projects the chance of failure is much greater because many of the key personnel in the system are actually willing that failure. In these circumstances the climate of implementation, the timescales and the level of ambition are all very different to projects undertaken in response to positive triggers. The analyst has to tread more carefully, take more time and be less ambitious in the proposed changes.

Such projects are very common. The problems they present are too fundamental to be solved by rational argument or advisory meetings and presentations. The analyst has no option but to get on with the system under conditions and constraints that may, at some time, seem to be intolerable. In such instances the computer is being used as a substitute for management and the analyst must recognise this and plan accordingly.

Neutral Triggers

"The insistence of the auditors that we improve our controls and reporting systems finally convinced us of our need to computerise".

Neutral triggers are events that are outside the scope and control of the system. These may be as a result of Government instructions (a need to present accounts in a certain way), legal requirements or the advice of external agencies such as professional associations or auditors. These triggers have little positive or negative effect on the project in hand.

The reasons behind making the large, and sometimes traumatic, step to computerisation undoubtedly affects the rest of the project. Thus it is necessary to probe and understand these reasons to help gauge the climate of the project and to adjust timescales and objectives accordingly.

Fortunately, an approach to project selection that recognises the strategic importance of information systems, should reduce the time needed to dwell on understanding the system triggers. The justification for the project has been discussed and debated at an earlier stage and at a more senior level, and so by the time it comes to implementation the need for the project should be well-documented and understood. This is not to say that all problems will be avoided, but many of the details of principle will have been sorted out. Participants should be aware of the *why*, *how* and *when* of the project and its relationship to the rest of the enterprise's plans and activities.

2.6 SUMMARY

This chapter has reviewed some of the issues that will be encountered early in a systems project.In summary it:

- Suggested that a computer system can help the doers and makers of an organisation obtain benefits but that it cannot obtain such benefits in itself. This led to a consideration of how information systems can be viewed in the perspective of business activity and objectives.

- Illustrated a 'top-down' approach showing how very broad objectives can be sharpened into specific goals that have information system consequences. It was suggested that these goals should be described in the context of the business (i.e. not have computer and non-computer bits) and that they should be quantified and time constrained.

- Identified ways in which IS might contribute to setting and achieving business objectives.

- Recognised the clinical precision of 'top-down' Strategic Planning and introduced two complementary approaches which acknowledged the rather fuzzy nature of organisations, problems and problem solving.

- Acknowledged that triggers of change are usually explicit in companies that have a strategic approach to planning and information system development. This is not necessarily true for enterprises that practise a more piecemeal development. In such instances it is important for the analyst to probe the reasons behind the desire for change and to recognise the implications of these reasons. The ambition, scope, approach and timescale of the project can then be judged accordingly.

Once the strategic importance of a project has been agreed, it can be subjected to a formal Feasibility Study.

3 The Feasibility Study

3.1 INTRODUCTION

Once a preliminary area of application has been established, it may then be subjected to a more rigorous examination in a Feasibility Study. The analyst will, of course, already have formed some ideas about the application from the preliminary planning work. However, the Feasibility Study represents an opportunity to 'firm up' knowledge of the system and to form ideas about the scope and costs of possible solutions. In many respects the Feasibility Study is a quick and dirty mini-Systems Analysis with the analyst being concerned with many of the issues and using many of the techniques required in later detailed work and covered in the following chapters of this book.

Feasibility Studies are usually undertaken within tight time constraints and normally culminate in a written and oral Feasibility Report. The content and recommendations of such a study will be used as a basis for deciding whether to proceed, postpone or cancel the project. Thus, since the Feasibility Study may lead to the commitment of large resources, it is important that it is conducted competently and that no fundamental errors of judgement are made. This is easier said than done.

This chapter examines three different aspects of feasibility and the compromise that eventually has to be made between them. The results of this compromise are eventually presented in a Feasibility Report which recommends whether a project should progress to the detailed investigation stage.

3.2 THREE TYPES OF FEASIBILITY

In the conduct of the Feasibility Study the analyst will usually consider three distinct, but inter-related types of feasibility.

3.2.1 Technical Feasibility

This is concerned with specifying equipment and software that will successfully support the tasks required. The technical needs of systems will vary considerably, but might include:

— The facility to produce outputs in a given time scale. For example, 20,000 examination certificates in three weeks.

25

— The ability to provide certain response times under certain conditions. For example, no more than a two-second response time at each terminal when there are four terminals being used simultaneously.

— The facility to input a large number of documents in a limited time scale. For example, 400,000 gas readings in one day.

— The ability to process a certain volume of transactions at a certain speed. For example, to report on seat availability and record airline reservations without a significant delay to the passenger.

— The facility to communicate data to distant locations, for example, Regional sales figures transmitted to an American Head Office.

In examining technical feasibility it is the configuration of the system that is initially more important than the actual hardware make. The configuration should show the system's requirements: how many workstations are provided, how these units should be able to operate and communicate, what input and output speeds should be achieved and at what print or screen quality. This can be used as a basis for the tender document against which dealers and manufacturers can later make their equipment bids. Specific hardware and software products can then be evaluated in the context of the logical needs.

At the feasibility stage it is possible that two or three different configurations will be pursued which satisfy the key technical requirements but which represent different levels of ambition and cost. Investigation of these technical alternatives can be aided by approaching a range of suppliers for preliminary discussions. The technical performance and costs of rejected alternatives should be documented in the Feasibility Report.

3.2.2 Operational Feasibility

Operational Feasibility is concerned with human, organisational and political aspects. General impressions of these factors may be gained from the corporate appraisal and through consideration of the system trigger. Amongst the issues examined are:

— What job changes will the system bring? Most people react unfavourably to change. Planned job changes must be carefully handled so that those affected are seen to gain in a way that they feel is acceptable. This may be through job enrichment or simply through raising wages.

— What organisational structures are disturbed? The suggested system may cut across accepted organisational relationships and threaten the status of individuals and their promotional expectations.

— What new skills will be required? Do the current employees possess these skills? If not, can they learn them? How long will they take to learn?

It is unlikely that a project will be rejected solely on the grounds of operational infeasibility but such considerations are likely to critically affect the nature and scope of the eventual recommendations.

It should also be recognised that although the computer brings significant alterations to an organisation it is not the only source of change. The way that an organisation manages and implements changes in other areas should be considered when computer-related changes are planned. Agreed procedures for discussing proposals; Staff Consultative Committee, Trade Union Agreements, Staff Forums, etc, should be identified and followed.

3.3 ECONOMIC FEASIBILITY

Many organisations evaluate projects on an economic basis – they must show financial returns that outweigh the costs. For this reason, management tend to give more weight to economic feasibility than to technical and operational considerations. A number of approaches to assessing the costs of solutions have been suggested. Approaches include the following.

Least Cost

This is based on the principle that costs are easier to control and identify than revenues. Thus it assumes that there is no income change caused by the implementation of a new system or that two competing systems offer the same benefits. In such an evaluation only the costs are listed and the option with the lowest cost is selected.

Time to Payback

The 'Time to Payback' method of economic evaluation is an attempt to answer the question "How long will it be until we get our money back on this investment in systems?" This requires data on both costs and benefits. The net cash flow for each year is calculated by subtracting the value of benefits from the costs.

In the 'Time to Payback' method, the alternative which repays the initial investment the quickest is selected.

	PROJECT 1		PROJECT 2	
Year	Net Cash Flow	Cumulative Cash Flow	Net Cash Flow	Cumulative Cash Flow
0	–3000	–3000	–3000	–3000
1	1200	–1800	600	–2400
2	800	–1000	1000	–1400
3	1000	nil	1000	– 400
4	100	100	2000	1600
5	200	300	1000	2600

In the example given above Project 1 pays back in year 3 and Project 2 in year 4. Hence Project 1 would be selected despite its lower return in the long term.

This method of evaluation has two significant disadvantages.

- The example given above shows how the approach only considers the time taken to return the original investment and ignores the system's long term profitability. Thus options that are more profitable in the long run are not selected.

- The method does not recognise the time value of money. Benefits that accrue in the distant future are not worth as much as similar benefits which occur more quickly, but the 'Time to Payback' method fails to recognise this.

Net Present Value

This is a well-defined and practised method of economic evaluation. It builds in an allowance for the 'time' value of money, represented by the Present Value Factor. In this method the net cash flows are reduced in value by applying this factor, so reducing the value of a cash flow to its present worth. For example, a cash flow of £30,000 planned for ten years time will actually only be worth £4,800 if a Present Value Factor of 20 percent is used. Thus benefits that appear late in the project's life span contribute little to its economic feasibility. This may make project selection a little conservative. In many computer-based projects, positive cash flows appear late in the project life and heavy discounting of these benefits may lead to the selection of less ambitious projects which yield a quicker return.

For example:

A company is considering the computerisation of its stock system. The projected extra cash inflow from stock savings and improved order handling is as follows.

Year 1	2000
Year 2	7000
Year 3	7000
Year 4	4000
Year 5	3000

To achieve these benefits the company is considering the purchase of a computer system and associated software. This has an initial cost of £15,000 with annual payments of £1500 for maintenance.

A simplistic cash flow can be constructed ignoring the time value of money.

Year	Cash inflow	Cash outflow	Net cashflow
0	0	15000	−15000
1	2000	1500	500
2	7000	1500	5500
3	7000	1500	5500
4	4000	1500	2500
5	3000	0	3000
		Total	2000

This gives a total positive cash flow of £2000.

The use of present values introduces the principle of time preference. In this approach future cash flows are discounted using an appropriate interest rate. These discount factors can be calculated using an appropriate formula or taken from published Discount Tables.

The present value of cash flows using an interest rate of 10 percent is shown below:

Year	Net cash flow	Discount Factor	Present Value
0	-15000	1.0000	−15000.00
1	500	0.9091	454.55
2	5500	0.8264	4545.20
3	5500	0.7513	4169.71
4	2500	0.6830	1707.50
5	3000	0.6209	1862.70
		Total	−2260.34

The total present value (or Net Present Value – NPV) of the project is less than zero and this indicates that, other things being equal, the project is not worth doing.

The same approach can be used for comparing two investment options. The table below shows the purchase and lease options for a particular project.

	PURCHASE			LEASE	
Year	Cash Inflows	Cash Outflows	Net Cash Flows	Cash Outflows	Net Cash Flows
0	7500	10000	−2500	4000	3500
1	3000	2500	500	4000	−1000
2	3000	2500	500	4000	−1000
3	3000	2500	500	4000	−1000
4	3000	2500	500	4000	−1000
5	3000	0	3000	2000	1000
Total	22500	20000		22000	

At first sight the leasing option seems more expensive than the purchase alternative. However, both options produce a positive overall cash flow.

Present values for this example are shown in the following table. A 10 percent interest rate is again used.

		PURCHASE		LEASE	
Year	Discount Factor	Cash Flow	Present Value	Cash Flow	Present Value
0	1.0000	−2500	−2500.00	3500	3500.00
1	0.9091	500	454.55	−1000	−909.10
2	0.8264	500	413.20	−1000	−826.40
3	0.7513	500	375.65	−1000	−751.30
4	0.6830	500	341.50	−1000	−683.00
5	0.6209	3000	1862.70	1000	620.90
			947.60		951.10

In this case both options almost produce the same net present value. Both are positive and this indicates that, other things being equal, the project is worth doing.

The Internal Rate of Return (IRR) is a related measure. The IRR is the discount rate which yields an NPV of zero for the project. This IRR can then be compared with the cost of capital (the interest rate) to see if the project will be profitable.

Breakeven Analysis

This technique is particularly useful when the system is subject to varying workloads. It distinguishes between fixed and variable costs and fixed and variable benefits. The data is plotted on a graph where the vertical axis is the amount of cost or benefit, and the horizontal axis is the increasing level of the workload. The fixed costs are plotted first with the variable costs plotted above them to show the increase in total costs as the workload increases. The same is done for the benefits. The crossing point of the total benefits line with the total costs line indicates the breakeven point. Workloads to the left of this point do not justify the use of the system, while workloads to the right of the breakeven point do. Breakeven analysis does not give the full picture needed for the economic evaluation of systems, but its emphasis on the operational phases can be useful.

There are a variety of methods which may be used in economic evaluation. These methods may give contradictory advice and none of them enjoys universal acceptance. However, whichever economic evaluation method is adopted (except the rather simplistic Least Cost) there will be a need to predict and quantify benefits. This is typically more difficult than quantifying costs. Chapin (1981) identifies several reasons for this:

— Uncertainty about the timing and amount.

— Problems of expressing certain benefits in direct monetary terms. What is the value of not having to apologise for as many order errors? Many benefits will often appear as 'intangibles' – better management information, improved management controls, etc.

— The benefit is often due to a joint effort of a number of departments in the organisation. In such cases it is very difficult to assess the computer's contribution.

You may recall that this latter issue was explicitly recognised by suggesting that all costs and benefits incurred in a system project should be included, not just the computer system's contribution. This is a further justification for taking a 'top-down' approach to project selection through seeking business objectives. The advantage not only lies in the inclusion of all costs but, more important, the benefit itself is defined by the goal. The significance of this must not be missed. Many conventional projects are subject to a set of costs which must be offset by scrabbling around for presumed benefits. These costs are often set in an inappropriate way – someone 'decides' that the budget will be £10,000, because this 'sounds a reasonable figure'. All subsequent costs and benefits have to be squeezed into this framework. Benefits may be hard to find, so intangible gains such as better control, up-to-date reports, etc., are introduced to make the project look financially attractive.

However, if the alternative perspective of defining benefits before costs is adopted, then the framework for system development immediately becomes more realistic. For example, the overall ambition of the project should become clearer. The budget for computer hardware and software can be set within the benefits that are likely to accrue. If savings of £250,000 per annum are envisaged then it is probably unrealistic to restrict thinking to £10,000 systems. A recognition of this will permit more innovative technical solutions and operational strategies. Many years ago I worked on a project blighted by the Managing Director's insistence of an £8,000 budget. It was not possible to develop the system for this money and the project was subsequently dropped. It was later pointed out that development at this price would permit the system to payback in six weeks, a performance not matched (or required) by any of the company's other investments. The Managing Director acknowledged this but was unmoved. A consultant had told him that a system could be bought for the price of an average saloon car and he had no reason to doubt that advice.

3.4   THE FEASIBILITY REPORT

The Feasibility Study will usually culminate in a formal written report and an oral presentation. The possible contents of a Feasibility Report are listed below.

Introduction
Background to the project. A brief review of the layout of the presented report.

Terms of reference
This is likely to include reference back to the preliminary analysis and an explanation of how the system under discussion was selected as a candidate for investigation. It will also include details about the scope, resources, timescale and client of the Study. It is important that these are established and agreed at the outset. Possible terms of reference for a sales support project might be:

> To undertake a Feasibility Study of the sales support function identified in the Corporate Strategic Plan (*see* Section 2.5). The Study will commence on the 1st August 1993 and be undertaken by one Senior and two Junior analysts. A formal report will be presented to members of the Information Systems Planning Group on the 11th November 1993, followed by an oral presentation at the Group meeting on the 24th November. The Feasibility Report will adhere to company standards (*see* ref 45/23/65) and evaluation criteria. The client for this project is Keith Freeman, Sales Manager – Northern Division.

Terms of reference should be explicit about what is expected, by when and what resources are available to achieve this. The intended outcome from this project stage is particularly important. Many projects are blighted by unfulfilled expectations.

"We expected to see a system not a report!'

The term deliverables is often used to describe these outcomes. Terms of reference should always include the intended deliverables. In the example given above, these are a report and a presentation. In other instances they might be a system, a program, a memorandum, etc.

Existing System

A description of the relevant system(s) currently operating in the organisation. These will have be investigated using the fact finding techniques described in the next chapter and presented using appropriate methods such as Data Flow Diagrams and Logical Data Structures. These will be less complete than in the subsequent detailed analysis, but special attention must be paid to the technical requirements. Particular problems will be highlighted and the implications of these discussed.

System Requirements

These will be derived from the existing system (outputs currently produced may still be required when the system is replaced) and from discussion with system users and operators who have identified requirements that are not presently fulfilled. Critical performance factors must also be covered, (e.g. the need to produce 5,000 invoices per day, to process transactions in less than 5 seconds) because these will have an important bearing on the hardware selection. Audit, Security and Data Protection implications may also be discussed.

Proposed logical system

A number of business systems are developed to reflect different combinations of requirements. One of these options is selected by the project client and a required logical Data Flow Diagram and Logical Data Structure developed to reflect the agreed business scope. This outline logical system design may be supported by definitions of key inputs and outputs described more in their content than in layout and display.

The differences and advantages of the proposed system over its predecessor will be highlighted, together with its effect on other systems currently operating in the organisation. The new system may impose certain constraints in operation (e.g. all input documents must be submitted by 4 p.m.) and these should be clearly described and discussed.

The possible effects on staff must be identified and a strategy for staff training, reduction or redeployment suggested or requested.

Proposed physical system

A number of outline technical solutions may be evaluated and a preliminary technical implementation agreed.

The extent to which this can be done depends upon the current resources of the organisation. If the firm already has a large computer then the extra hardware is likely to be additional terminals, more secondary storage, perhaps more memory. It is likely that these will not be required until the system becomes operational and so the specification may be altered as detailed analysis and development clarifies the nature of the system.

In contrast, large projects and organisations without significant computing resources will have to invest in hardware before development can get under way. As a result there are significant pressures to select and purchase the hardware very early in the project's life, before many of the detailed implications of the system have been discovered.

Development Plan

This will define a suggested project definition and plan for the detailed analysis and design phases which will follow the acceptance of the Feasibility Report. A complementary plan will also be needed for hardware and software evaluation, purchase and installation. In effect, the Feasibility Report establishes the terms of reference for succeeding projects.

Costs and Benefits

These have already been discussed. They will clearly vary in detail and accuracy, as will the techniques used to evaluate them.

Alternatives Considered

In the process of arriving at a suggested system the analyst usually considers and rejects a number of options. It is important to record these considerations for two main reasons. First, it may nip a number of time-consuming "have you considered....." discussions in the bud. Secondly, it permits the sponsor of the Study to examine the legitimacy of the reasons for rejection. For example, the analyst may have rejected a certain option because, in his impressions gained from the preliminary analysis, it appeared to be too costly. However, the information contained in the Study may now pursuade the sponsor to change his mind about the level of ambition of the project and so the rejected alternative becomes feasible. This alteration would be unlikely if details of rejected alternatives were not included.

The report would normally end with conclusions, recommendations and relevant appendices.

In general the recommendation will be

- to proceed with the project;

- to abort the project and invest elsewhere;

- to review the scope of the project and hence re-define costs and benefits until it becomes justifiable.

The Feasibility Compromise

The three ways of approaching feasibility are likely to conflict. In general, 'better' technical solutions cost more money, while robust, helpful, user-friendly software is time consuming to write and hence incurs high development costs. Such software may also mean larger programs and so the system has to carry a much larger software overhead, and this may begin to conflict with performance requirements. In many instances technical and economic factors become paramount – "the system must have a two second response time and return its investment in three years" – and so the operational factors become devalued. This often has unfortunate consequences.

The Feasibility Study differs from analysis 'proper' in its level of detail. It is difficult to give general advice on what constitutes an acceptable depth of analysis because this will vary with the organisation and the application. There is always the nagging worry that the detailed analysis work will uncover a hitherto overlooked fact that now makes the project infeasible. This is further complicated by the difficulty of reconciling the three feasibility criteria, particularly with the insistence of many organisations on an economic cost benefit analysis. The restricted timescale of a Feasibility Study also makes it difficult to comprehensively evaluate and offer sufficient options at different levels of cost and ambition.

For microcomputer applications there are special difficulties in conducting Feasibility Studies. In large organisations the task may be given to a Senior Analyst, but in a firm computerising for the first time there is no equivalent person. Thus the enterprise is very dependent upon its own non-specialist staff and the integrity of possible suppliers. In most instances the suppliers of microcomuters will not have the necessary resources, skill or time to perform a proper Feasibility Study. It is difficult to justify even one day of analysis on a job with a likely profit margin of less than £1000. Thus the preparation for computerisation may be less than ideal.

Prototyping may have an important role to play in these early stages of the project. Two examples must suffice:

— In one instance the analyst was concerned about meeting the technical output requirements identified in preliminary discussions with the main user. If these could not be attained using relatively simple technology then the cost of purchasing and maintaining a new, advanced printer would render the project economically infeasible. This problem was resolved by creating a set of test programs that produced and timed representative output. The quality and speed of the printing was checked with the user who agreed that they met his requirements. This impressed the user as well as eliminating one of the analyst's doubts.

— In another project the analyst was faced by a set of operators who had already suffered from a poorly planned computer installation. They seemed sceptical of his plans until he used a commercially available package to demonstrate the opportunities that existed. The operators made detailed criticisms, but their attitude towards the whole project became more positive as the session progressed. This simple idea changed aspects of the Operational Feasibility within one hour.

3.5 SUMMARY

Once possible areas of application have been identified they should be subjected to a Feasibility Study. This chapter has:

- Identified and described three feasibility criteria – technical, operational and economic.

- Introduced four approaches to economic feasibility.

- Outlined the contents of a Feasibility Report.

- Recognised the compromise between the three feasibility criteria and the possible role of prototyping in arriving at that compromise.

The Feasibility Study will culminate in the Feasibility Report which will be presented to management for approval. If this approval is forthcoming then detailed analysis and design will commence.

4 Fact Gathering

4.1 INTRODUCTION

Analysis is primarily concerned with three tasks.

- Finding facts that will permit understanding of the present system and aid the design of any successor.

- Mastering fact finding techniques that enable the finding of these facts.

- Organising the facts into a rigorous set of documentation.

This chapter concentrates on the middle issue: fact finding techniques.

However, before embarking on an examination of these fact gathering techniques, it is useful to briefly discuss the other two aspects of this trio so that the context of this chapter can be better understood.

In many respects most of this book is about what facts to find. The models that are constructed in this and subsequent chapters demand information that must be uncovered. Thus, for example, the Corporate Appraisal (*see* Chapter 2) requires facts about the economic structure of the industry, the competitive position of the company and the climate of Industrial Relations. The System Flowchart (Chapter 5) will demand the uncovering of administrative arrangements, information flows between sections and the location of files. Data Flow Diagrams and their supporting Data Dictionary entries (Chapters 7 and 9) will mean that logical processes have to be specified in detail. The same **investigative** nature is true of Entity models (Chapter 6), their very construction will raise problems that have to be resolved through subsequent fact finding.

Two points must be stressed:

- Facts will not emerge in neat bundles. General issues about the department are likely to be wrapped up in administrative detail and discussions of office politics.

- It is impossible to ever be sure that fact finding is complete. Partitioning of analysis and design is likely to be counter-productive. Facts will continue to emerge during the 'design' stage.

It is the fact recording techniques that prompt what facts to find and give the protocols for modelling them. Thus Documentation Standards attempt to direct fact recording by requiring the completion of a standard set of diagrams and documents. The role of these standards is introduced in this chapter but examples of their use are found throughout the book. For example, a Data Flow Diagram is constructed with a standard set of symbols and rules which are introduced when the construction and purpose of this model is considered.

4.2   FACT GATHERING TECHNIQUES

4.2.1   Background Reading

Organisations usually amass a considerable amount of documentary evidence and this can provide the analyst with an important insight into current organisational arrangements. This evidence may be available in many forms, although it must be recognised that smaller companies may be less well formalised or documented because of time and resource pressures. Reports of previous market surveys or feasibility studies may be outdated but worth reading to understand the background to the current study. Company information may be available in the form of:

— Organisation charts.

— Adminstrative procedure manuals.

— Job descriptions and specifications.

— Training manuals and memoranda.

— Sales literature.

The Strategic Plan and its associated Corporate Appraisal will also be significant sources of relevant information.

4.2.2   Interviewing

Interviews are formal meetings where the analyst can obtain information about the operations of the present systems and the requirements of any planned replacement. Successful interviewing is a skill that can be developed through practice. Furthermore, it is undoubtably the most common fact finding method (see for example, Sumner and Sitek, 1986) and has to be conducted with the full cooperation of all employees. Successful interviewing is particularly important in the earlier stages of a project. Many people place a high value on first impressions and their cooperation in the development of a new system will be hindered if the initial contact is unsuccessful.

Interviewing will be used at various times and for different purposes as the project progresses. For example,

— to gather facts about the procedures and decisions taking place in an organisation;

— to check the analyst's understanding of system operations with users of all levels;

— to validate aspects of a proposed system design;

— to build confidence in the design of a new information system.

The purpose of the interview will determine the balance of the discussion between the interviewer and the interviewee. In general, most of the interviews in the early parts of the project will require the analyst to **listen**, the latter ones to **explain**.

4.2.3   Planning an Interview or Discussion

Planning is essential and may take as much time as the conduct of the actual interview. Three aspects of the proposed interview need to be planned.

— The objective or purpose of the interview

It is essential to decide what will be accomplished by the discussion. Wherever possible this should be expressed in terms of targets, such as 'to identify the store manager's view of the stock control problems' or 'to define the structure and format of the monthly stock report'. The purpose of the meeting should be clearly explained to all participants so that relevant documents and information can be collected in advance. An agenda is an important method of formalising the meeting objectives as well as identifying documents and facts that can be established prior to the meeting . The agenda also tends to raise the status of the meeting and emphasises that this is not an informal conversation. An example agenda is given in Figure 4.1.

— The time and venue of the interview

A mutually convenient time, duration and place for the meeting must be established beforehand as this will allow the participants to schedule work accordingly. It may be preferable to be away from the normal workplace if this guarantees privacy from interruption. However, this will mean that records and examples will not be to hand.

— The authorisation for the interview.

The analyst should always obtain the appropriate manager's consent before interviewing staff. Where a series of interviews is required then it is sensible to work downwards through the organisation's hierarchy. Knowledge of the interviewee's position and duties and preparation on the subject for discussion helps generate confidence. To eliminate the possibility of arguments or differences of opinion it is wise to interview only one person at a time. Insecurity can sometimes prompt a manager to request attending a subordinate's discussion and the "one-person" rule can be used to deny this. It is important to gain independent views of the system. There is often a large gap between how the manager thinks the system works and how it actually works.

4.2.4   Conducting the Interview

A clear introduction and explanation of the purpose of the interview should be given. If an interviewee is left unsure of the identity and role of the interviewer then the freedom of the discussion will be restricted and the interview prove less effective.

Elementary good manners should be observed. For example, being punctual and keeping the discussion to the time agreed in advance. Conforming to certain dress requirements is a further example. There are enough problems in conducting good interviews without erecting avoidable barriers.

ADAPT : Meeting Agenda			System *Example*		Ref: MA *21*
Author *SD*	Date *01/01/92*	Meeting Date *03/01/92*	Meeting Time *13.00*	Duration *45 mins*	Page *1* of *1*

Participants

Mr. T. Conner - Head of Purchasing

Mr. S. Dalton - Senior Systems Analyst

Comments

Agenda

1: Terms of reference for the project

2: Overall organisation structure of the purchasing department.

3: Overview of the functions of the purchasing department.

4: Any other business

Documents to be made available:

- organisation chart

- job descriptions for each employee grade

Figure 4.1 An example agenda

Many interviews do not meet their objectives simply because they are poorly controlled by the analyst. This does not mean that the analyst should slavishly follow a pre-prepared checklist and abruptly steer the interview in an uncompromising manner. Some gentle re-direction of the discussion is usually enough to bring the interview back to its immediate aims. Some care also has to be taken in imposing such control. Permitting the participants to develop certain themes can give very useful insights into departmental tensions and operational difficulties. This information may ultimately prove more useful than the points marked for discussion in the analyst's pre-interview checklist. Thus a willingness to replan an interview is vital if unexpected information comes to light.

The questions asked should always be relevant to the interviewee and appropriate to their knowledge and status. It is always advisable to use the interviewee's language to describe an aspect of a system and not to introduce unnecessary jargon from the information systems world (interface is a particular word to avoid). During the interview the analyst should try to identify the following types of inadequate response.

— *Non-response.* The interviewee refuses to answer a question. The very act of refusal may give an important insight. An accountant's refusal to discuss his relationship with the marketing department may indicate certain inter-departmental conflict.

— *Inaccurate response.* This may occur through deliberate or accidental distortion. It is very common to find that two or three members of an organisation give slightly different descriptions of how certain operations are undertaken. It has already been recognised that there is often a distinction between how a manager thinks something is carried out and how it actually is. All responses should be cross-checked.

— *Irrelevant response.* The interviewee does not give an answer to the question asked. This tends to be a waste of time but is little more than an irritation if the information can be elicited from the next, more carefully phrased, question.

— *Inappropriate question.* The interviewee lacks the necessary information to frame an adequate response. This gives some ideas about the organisational boundaries. Incorrect information about part of a certain system may suggest that the interviewee has little direct connection or involvement in that particular operation.

— *Partial response.* A relevant but incomplete description is given. This is particularly dangerous as the analyst must be able to design a system that covers all forseeable possibilities.

A successful interview is likely to be one in which the interviewer listens a great proportion of the time, giving the interviewees the opportunity to express themselves in an unrestricted manner. As a result the planned discussion area will be comprehensively, but unobtrusively, covered. During the interview the analyst must also be aware of the body language of the participant. Gestures, facial expressions, degree of eye contact and general posture of the interviewee will convey a great amount of information about values and opinions. These will require responses on the part of the analyst. For example

- Showing interest and warmth.

- Providing encouragement by nodding, smiling or eye contact.

- Asking questions if the interviewee appears to falter.

Analysis interviews are concerned with communication. This demands that the analyst gives the participant both the opportunity and encouragement to express facts, opinions and fears. Communication is a social activity – one way communication is not true communication (Cherry 1978).

4.2.5   Concluding the Interview

Keeping to the agreed time is important and it is usually desirable to arrange a further session rather than to extend the current one. Both participants will have scheduled subsequent tasks and concentration will deteriorate as they become aware of these. The interview should conclude with a brief résumé of the discussion and the agreement of all action points.

The analyst will need to write up the discussion points promptly, clearly and, where points of detail or policy are involved, to send a copy to the interviewee for approval. These minutes should help clarify points of misunderstanding or omission, as well as identifying responsibilities for further action.

4.2.5   Questionnaires

When it is impossible because of time, distance or simply cost, to interview all the desired people involved in a system then the analyst may consider the use of a questionnaire. This is a more structured and formal method of collecting data, but may be the only viable option where there are a large number of dispersed users.

Designing a series of questions which seek out the required information is a skilled activity and careful planning is required if the results are to prove of any value. There will have to be a balance of open-ended and closed questions. Open-ended questions simply ask a question and leave an adequate space for an unstructured response. The replies may give important insights but they are very difficult to analyse in any systematic or statistical way. An example of an open ended question is given in Figure 4.2.

What do you consider to be the most significant problems you face in the operation of the current booking procedures?

Figure 4.2   An open ended question

In contrast, a closed question is useful for eliciting factual information. This demands that the analyst has a good understanding of the area under review so that simple effective questions can be formulated. Each question will have a clear purpose and be posed in an unambiguous way, allowing the respondent to give a definite answer. This may be through a selection of answers, actual examples or on a scale of agreement or disagreement. An example of a closed question is given in Figure 4.3

How many Full-Time Sales Staff are currently employed in your Sales Region?

1. less than 10
2. 10–29
3. 30–100
4. More than 100

Figure 4.3   A closed question

Unfortunately questionnaires have several drawbacks which contribute to their under-use in the investigation of information systems. Low response rates to the questionnaire may lead to unrepresentative responses being analysed to draw incorrect conclusions. Lack of direct contact with the analyst may mean that questions are interpreted in different ways. There is no opportunity to clarify ambiguities unless a follow-up visit or phone call is made. Finally, there is no possibility of the analyst observing the user's workplace or work practices. The richness of information imparted in the body language of the interview is forsaken.

A few guidelines in questionnaire design are offered below.

- Clearly explain the purpose of the questionnaire.

- Phrase the questions so that they are unambiguous, concise, and unbiased.

- Avoid long questionnaires with many different types of information being sought.

- Test the wording and structure of the questionnaire and ask sample respondents to explain their understanding of each question.

- Decide how the results are to be analysed beforehand. A questionnaire can be set out in a way which permits direct input of answers into a computer system, where results can be analysed with an appropriate statistical package.

- Impose a deadline for the response and include a prepaid envelope for postal responses.

It is important that questionnaires are not completely dismissed as a fact-gathering technique. They can be used to 'weed out' a large number of possible respondents so that fact-finding interviews can be better focussed. Furthermore, they also have an important role to play in the selection of reference sites, suppliers and the evaluation of software. Significant advantages include:

— It is relatively cheap, particularly when there is a scattered group of users and operators.

— It is free of interviewer distortion and error.

— It permits time to refer to documents and documentation. Questions which concern detailed factual data – e.g. How many customers live in the South West? are suited to questionnaire collection.

— It may be possible to ask more personal and controversial questions particularly if the response is to be anonymous.

A fuller discussion of the relative merits of interviews and questionnaires is given in Moser and Kalton (Moser and Kalton 1971).

4.2.7   Analysing Documents

Documents are very important because they represent the formal information flows of the present system. During all discussions the documents used by the participants will be referred to, examined and explained. The analyst should collect specimens of all these relevant documents – input forms, output analyses, reports, invoices, etc., in an attempt to understand how data is passed and used in the present system.

Each document will have its own cycle of creation, amendment, use and deletion. Relevant questions might include:

— What event triggers the generation of the document?

— Who generates the document?

— How is it prepared?

— Where is the data derived from?

— Who uses the document?

— For what purpose is it used?

— How is it stored?

— How long is it kept for?

For example, a timesheet recording the hours worked by an employee will be generated every week by that employee, verified by the foreman, used to calculate the wages and to allocate costs to specific projects, and then filed.

The details that are collected about the document will be guided by the appropriate Documentation Standards (see later) but features might include:

— when each data item is entered;

— the meaning, size and format of each data item;

— the source of each data item;

— the use of each data item;

— the filing sequence.

Document analysis should also include an assessment of the clarity of the form and how well it satisfies its purpose. The analyst is particularly looking for ambiguity or obsolescence; for example, column headings that do not correctly indicate the data entered under that item. The volume of documents produced or received is also significant. Any new system must be able to cope with the amount of data passing through it including seasonal peaks or other variations. A simple average measure is of little use in these circumstances as a system built around averages will underperform for almost half of the time! The growth rate of document use is also important. If the number of documents (say invoices) is increasing by 12 percent per annum then it would be inappropriate to tailor the system around current volumes.

	Request particulars	Arrange viewing	Contracts matters	General enquiry	Others
Monday					
09.00–11.00	‖‖	‖		‖‖	‖
11.00–13.00	‖‖	‖	‖	‖‖	‖‖
13.00–15.00	‖‖‖	‖			
15.00–17.00	‖‖	‖	‖	‖‖	‖
Tuesday					
09.00–11.00					
11.00–13.00					
13.00–15.00					
15.00–17.00					
Wednesday					
09.00–11.00					
11.00–13.00					
13.00–15.00					
15.00–17.00					
Thursday					
09.00–11.00					
11.00–13.00					
13.00–15.00					
15.00–17.00					
Friday					
09.00–11.00					
11.00–13.00					
13.00–15.00					
15.00–17.00					
Saturday					
09.00–11.00					
11.00–13.00					
13.00–15.00					
15.00–17.00					
Sunday					
11.00–13.00					
13.00–15.00					

Figure 4.4   Special Purpose Record

Documents may also be copied – or be multi-part in the first place – with copies going to different people or departments. Each copy must be traced, its purpose ascertained and the filing sequence noted as this is likely to indicate how old copies are currently retrieved. For example, if copy invoices are filed in date order then it is very likely that their retrieval will be based on date of invoice, not by any other data item on the invoice such as customer name. This information may be of importance in the subsequent design of files on any proposed computer system. We may decide to index on date of order as well as on more likely fields such as customer name or account number.

4.2.8   Special Purpose Record Keeping

Some information which the analyst requires cannot be obtained by directly interviewing participants or collecting documents used in the present system. This may typically be quantitative aspects such as volumes, trends, frequencies and time intervals. For example, the analyst may wish to know the relative frequency of different types of enquiry dealt with by a receptionist. It is unlikely that such information would be routinely available and so a special purpose document would be designed to collect it. Figure 4.4 shows an example of the tally chart kept on a receptionist's desk to record the enquiries from customers.

The most effective special purpose records are those which do not impose an administrative burden on the person required to collect the data. A simple chart requiring a single mark as each event occurs will be completed more accurately than a complex form or instrument which the operator does not fully understand. It must also be recognised that the records refer to a sample of events, operators or frequencies and so every attempt should be made to select a representative sample.

Specific opportunities for special record keeping arise when the system is currently computerised. For example, the present pattern of report requests can be logged by writing simple audit programs that collect certain data about the enquiry (such as date, time, who requested it, what was requested, report generation time, etc.) but do not interfere with the report enquiry in any way. The pattern of use can then be analysed. Relevant information may also be routinely available from the System Log of the Operating System.

4.2.9   Observation

Observation of a system in normal operation will expose many features which might not be considered relevant, discussed or documented in any formal way. This may prove useful in gathering information on office conditions which the employees take for granted. A newcomer will notice environmental conditions such as levels of noise, lighting and interruptions. The analyst may be able to observe the normal levels of supervision and control, the flow of work, the occurrence of bottlenecks in the workflow, the pace of work and the levels of normal and peak workload. Informal systems for producing and storing information, such as personal data files or aids to working, are often only discovered when the analyst observes a person doing their job. Inter-office communication and handling of spontaneous queries are also more easily observed.

The value of observation as a fact finding technique depends upon how long the activity is undertaken and the skill of the observer. It is very time-consuming and so is best used to supplement other techniques in the building of a picture of the organisation and its information needs. Experience shows that the analyst may also disturb the work patterns that are being observed and this will naturally reduce the value of the fact finding exercise.

Finally, it must be acknowledged that observation may be the only feasible method of data collection in certain systems. For example, in a system to control and plan reservoir levels, the pattern of water inflow has to be measured and observed. In such instances, interviewing is clearly inappropriate.

4.3   DOCUMENTING THE FACTS

Facts gathered by interviewing, observation, document collection and other techniques will be used by the analyst in many subsequent tasks. Therefore, they need recording in some standard way for future reference by members of the development team. Standards for documentation may differ from organisation to organisation but the examples used below feature the basic elements which need to be recorded.

Standards have a wider use than just documentation. They should aid analysis and design by prompting the analyst to pose certain questions so that parts of the documentation may be completed. For example, the clerical document analysis form (Figure 4.6) requires the analyst to insert the maximum and minimum number of documents passing through the present system. Thus the analyst knows that such data must be obtained and so he may include the question in an interview with the relevant user. In this way documentation provides **prompts** for action not just a **record** of actions.

Standards also represent a method of controlling projects. The Head of the Systems section can effectively control the way that all system projects are developed by insisting upon completion and adherence to certain standards. This may be very useful because, as Daniels and Yeates once wrote, "until the system is implemented the only tangible evidence that the analyst has done any work is his documentation".(Daniels and Yeates, 1971).

Finally, Standards impose modelling tools, such as flowcharts, which permit the communication of facts between the analyst and the users and between the analyst and his colleagues and managers.

Thus standards are primarily concerned with **aiding** analysis and design, **documenting** the results of analysis and **communicating** these results to other people. The examples described below show documentation to support certain areas of fact finding.

4.3.1   Recording Discussions

A successful interview will trigger a seemingly continuous flow of information and the analyst will face the difficulty of recording it all. Much of the information which is obtained will be summarised and translated into models of the organisation or system. However, many aspects of a discussion cannot be forced into such a format without losing value. Opinions, worries and requirements may all be discussed in an interview and these need to be minuted to form a permanent record for future reference.

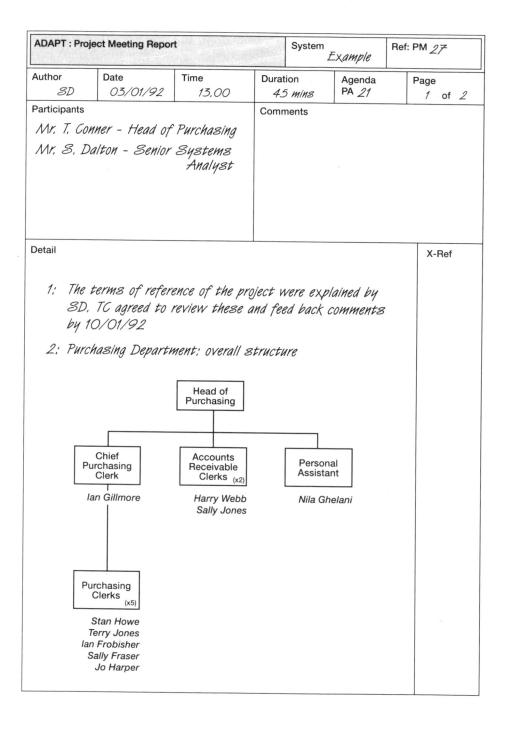

Figure 4.5   Discussion Record

ADAPT : Project Meeting Report			System *Example*		Ref: PM *27*
Author *SD*	Date *03/01/92*	Time *13.00*	Duration *45 mins*	Agenda PA	Page *2* of *2*

Participants

Mr. T. Conner - Head of Purchasing

Mr. S. Dalton - Senior Systems
Analyst

Comments

Detail

X-Ref

3: Job descriptions were provided for each grade of employee.

4: The purchasing clerks are responsible for stock purchase and for negotiating terms and discounts.
They report to the Chief Purchasing Clerk
Details of operational procedures will be explained by Ian Gillmore and an appointment was agreed for 07/01/92 at 9.00 hrs.

5: Purchasing clerks need to liaise with the accounts receivable clerks who monitor the financial implications of purchasing plans. These two clerks are also responsible for ensuring correct receipt of goods and authorising payment.

6: The possibility of word processing training for Ms Ghelani was raised. SD agreed to report back by 08/01/92

7: A meeting with Harry Webb and Sally Jones was set up for 10/01/92 at 9.30 hrs.

Figure 4.5(contd)   Discussion Record

The information will be required in many future analysis and design tasks and the analyst will need to refer back to past interviews as the models described in the rest of this text are developed. Thus a Data Flow Diagram (*see* Chapter 5) will be built up from documented knowledge, whilst in itself generating the need for further interviews to clarify points, explain issues and discuss details. Discussions and the **record** of those discussions underpins a considerable amount of analysis and design activity. A Discussion Record document (*see* Figure 4.5) can be used to summarise the meetings that take place as the project progresses.

It is important to stress that the discussion record is not a script. An understanding of the information which has been given is more important than the actual words. A summary in the form of a narrative or a model or picture of some kind is usually adequate, although the level of detail will depend on the purpose of the interview.

4.3.2   Recording Documents

A completed Document Analysis form for an example document is shown as Figure 4.6. Note how it prompts for the information needed to give the analyst a deeper understanding of the document's use and content. Document description cannot be said to be satisfactorily completed until all of the sections of this form are filled in.

This provides a comprehensive but static view of the data collected on the form. It can be supplemented by a variety of grid charts (see below).

4.3.3   Problem/Requirements List

During the fact-finding process users will specify problems currently experienced with the present system and define the requirements of its successor. These must be documented in a formal way. This is the role of the Problem/Requirements List (PRL). This list (Figure 4.7) has the following structure:

- System administration header information.
- Brief narrative of the problem or requirement.
- A unique reference number for identification.
- The identity of the user(s) specifying the problem or requirement.
- A cross reference to document analysis and associated models.

The priority of each problem or requirement is for negotiation amongst users and may eventually have to be allocated by a 'super-user' or 'owner' of the system. However, this statement of priority is important since it will determine which issues must be addressed first. This in turn will affect the project planning and the setting of milestones. The solutions section of the PRL will be filled in later in the systems development process.

4.3.4   Document/Department Grid

This shows the relationship between the documents and the departments that process them. Document names are entered along the top of the grid and department names

Delivery Note   Note #: **1** | *9121-36* |

KRONENHALLE BREWERY LIMITED

Delivery Date: **2** *20-5-91*

Product: **3** *Export Pilsner*

Type: **4** *Bulk*

Quantity: **5** *1*

Customer Name: **6** *Quick Canners Ltd*

Address: **7** *38 High Street,*

Melton Mowbray,

Leics.

Special Instructions: **8** *DEPOT Closed Between 1200 &*

1300 hrs.

Customer
Signature

Figure 4.6a   Example Document

ADAPT : Document Analysis						System *Example*		Ref: DA *4 5*

Document Name *Delivery Note*		Ref. No. —		Author *SD*		Date *06/01/92*		Page *1* of *1*

Preparation						Storage		
By	Method	Frequency	Medium	Size	Parts	Order	Location	Duration
Brewery	*Hand*	*30/shift*	*Paper*	*A4*	*3*	*Delete no*	*Traffic office*	*1 year*

Volumes				Use			
Max	Min	Average	Growth	Users		Frequencies	
50/shift	*30/shift*	*35/shift*	*nil*	*Traffic Clerks*		*each delivery note 1/shift*	

Purpose *Delivery Instruction*	Comments *3 part - set: White, Red, Blue*

Ref No	Item	Format	Frequency	Value	Data Source
1	*Delivery Note number*	*9999-99*	*1/note*	*–*	*Brewery*
2	*Delivery date*	*Date*	*1/note*	*Valid date*	*"*
3	*Product*	*Character*	*1/note*	*Any valid product*	*"*
4	*Type*	*Character*	*1/note*	*Bulk or Can*	*"*
5	*Quantity*	*999*	*1/note*	*0-999*	*"*
6	*Customer Name*	*Character*	*1/note*	*Any valid customer*	*"*
7	*Customer Address*	*Character*	*1/note*	*cust-adress for ⑥*	*"*
8	*Special Instructions*	*Character*	*1/note*		*Depot list*
				–	

Figure 4.6b   Document Analysis

ADAPT : Problem/Requirement List			System _Example_	Ref: PR _01_

Author _S.D._	Date _17/01/92_	User _Agent_	Priority _1_	Page _1_ of _74_

Problem/Requirement Description

Currently impossible to distinguish between two (or more) shipments of the same sales or purchase order.

DFD/Entity Ref	Document Ref	Other Ref
_17_	_Purchase Order_	

Solution Description

DFD/Entity Ref	Document Ref	Other Ref

Figure 4.7   Problem/Requirements List

down the left-hand side. The numerical codes (see Figure 4.8) in the intersecting squares show the sequence in which each document is processed by each department. Thus Delegate List (1) is handled first by the Administrative Assistant who passes it to the Accomodation Officer before it is sent to the Seminar Manager. Nobody seems to use Delegate List (2) at all, so perhaps it is redundant.

Document/Data Item grid

This grid is designed to show the duplication of data items on various documents (Figure 4.9). Document names are entered along the top of the grid and data item names down the left hand side.

This can be used to identify duplication (does the Delegate telephone number have to be on the Delegate List?) as well as the omission of documents and data items. Figure 4.9 contains many data items that are outputted (O) but not entered (I), thus fact finding cannot be complete. The source of all the I(nput) data must be tracked down in subsequent investigation.

4.4   OTHER ISSUES

Fact gathering is a set of skills that benefit from practice and experience and they should be reinforced throughout the learning of the whole development task. There is a growing tendency, albeit perhaps an unconscious one, to present systems development as the mastery of a set of increasingly complex diagramming models. Kimmerly made this point in a Datamation article.

> "Current practices in systems analysis have a major deficiency: they place excessive emphasis on the technical details and structured and mechanistic methods, and show a corresponding failure to give due regards to aesthetics, imagery and creativity in the process of systems analysis."

> (Kimmerly, 1984)

Many of the technical skills are very useful but they need to be supported by good inter-personal skills. Communication, negotiation, tact and confidence are all important assets. Confidence grows from practice and experience and hence the need to give the student analyst the opportunity to undertake interviews and presentations where mistakes and shortcomings can be identified and tackled.

In a survey (Rosensteel, 1981) to determine the importance of various job skills to systems analysts, the four highest ranked skills were writing, listening, speaking and interviewing. However as Parkin (Parkin, Thornton and Holley, 1987) pointed out, it pays to look deeply at what is meant by communication skills.

> "It would be a mistake to interpret this (communication skills) as verbal fluency or the 'gift of the gab'. For example, one person in our experience had an extreme stutter which made communication in the ordinary sense very difficult. Nevertheless, he was a highly valued analyst."

They conclude that knowledge of the user area is a crucial foundation for good communication and that an analyst who knows what he or she should find is likely to collect important facts even if their communication skills are poor.

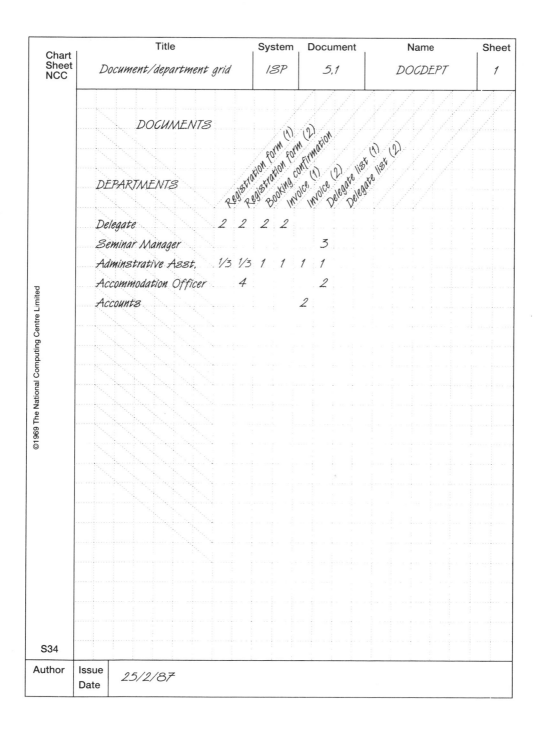

Figure 4.8   Document/Department Grid

Chart Sheet NCC	Title	System	Document	Name	Sheet
	Document/Data Item Grid	ISP	5-1	DOCITEM	1

DOCUMENTS

Registration Form
Booking Confirmation
Invoice
Delegate List
Conference Schedule

Data Items	Reg. Form	Booking Conf.	Invoice	Del. List	Conf. Sched.
Delegate-name	I	O	O	M	M
Correspondence-address	I	O	O	M	
Registration-date		O		M	
Delegate Title	I	O	O	M	M
Position	I			M	
Employer	I			M	
Tel-no	I			M	
Invoice-no			O		
Invoice date			O		
Invoice-net-amount			O		
VAT amount			O		
Invoice - gross amount			O		
Residential - charge			O*		
Residential - requirement	I*				
Non-residential - charge			O*		
Non-res-requirement	I*				
Meal - requirement	I		O		
Meal - charge			O		
B-B - requirement	I		O		
B-B - charge			O		
Cheque amount	I				
Seminar - title	I	O	O	M	M
Seminar - date	I	O	O	M	M

I = data item appears on document entering the system
O = data item appears on document when it leaves the system
M = data item contained on master record within the system
* = optional or mutually exclusive data items

S34

Author	Issue	
B. N.	Date	23/2/87

©1969 The National Computing Centre Limited

Figure 4.9   Document/Data Item Grid

They suggest that "good communication" occurs between the analyst and the user when:

- The analyst has acquired knowledge which the user's perceive as valuable.

- The analyst has an understanding of the user's environment and terminology which allows him or her to ask valuable questions and explain operations and requirements.

These are seen as the strategic requirements of good communication whilst taught skills such as interviewing techniques are conceived as a technical necessity. Thus the trainee analyst benefits from studying the business systems that underpin most technical activity.

Parkin et al conclude their discussion with the following hypotheses:

- An analyst who knows what he or she should find out is likely to get the important facts even if their communication skills are poor.

- An analyst who does not know what to find out is likely to fail to get important facts even if their communication skills are good.

The successful analyst must also be seen to undertake tasks in a fair and uncomplicated way. It is likely that he or she will be privy to information about current organisational performance, e.g. the 'inefficiency' of the stores department, yet the confidence of these 'inefficient operatives' must be gained to help analyse the problems of the current system. Furthermore, the analyst may also be expected to design a new system which makes these 'operatives' largely redundant. This presents dilemmas which must be appreciated but not resolved. These are matters of business policy which must be determined by line management, not systems analysts.

A further difficulty in fact-finding is that it is essentially looking for two kinds of fact. The first type exists in the current system and is typically discovered on present documents and in existing procedures. It may be difficult to uncover all the intricacies of the current system but at least they actually exist and can be recognised by the users and operators. In contrast, the second type of fact covers such areas as 'information needs beyond the present system', 'management reporting requirements' and 'management information'. In some instances these may be easy to identify, but in most circumstances it is impossible for users to envisage what might be useful or feasible in advance of receiving it. In these areas a prototype system can be used to experiment and monitor requirements. This might take the form of an application package because it is easier to criticise ("I see no use for that feature") and evaluate ("I did not think that such a report would be possible") a tangible product, than conjure up requirements 'out of thin air'. Groner (Groner et al, 1979) gives a case study example where

> "Prototypes were required in the requirements analysis phase because users could not be sure that computer systems were needed, what functions they should perform, or how they would use them."

Attempts to automate the collection of facts dates back to the reporting of AUTOSATE, developed by the RAND Corporation (USA), in 1964 (Autosate, 1964), but little reported since. Another early example was DATAFLOW where the fact recording was supposed to be done by relatively unskilled clerks using a specification language called DATAWRITE. This project was abandoned after unpromising field trials. In a survey (Parkin et al, 1987) it is suggested that fact finding is a worthwhile activity to automate but that few tools actually addressed this area. Certainly the time spent on this activity, and its perceived importance to both analyst and user seems to demand that automation is given greater attention.

4.5 SUMMARY

This chapter has reflected on the activity of fact gathering or fact collection. A number of specific methods have been introduced and examples given. The final section has introduced a number of facets of fact collection which require further discussion and investigation.

5 Fact Modelling

5.1 INTRODUCTION

This chapter examines two models for describing the current system under investigation. Both models focus on the passage of data through the system. The System Flowchart is an established technique that highlights the flow of data between departments, sections or personnel. It is well structured and easy to interpret.

The second model, the Data Flow Diagram (DFD), is used by SSADM to describe the current system and is later developed into its current logical equivalent (*see* Chapter 7). Unlike the System Flowchart it is not an end in itself but the first stage of a requirements definition model which will use the same notation and conventions.

Some analysts have discarded the System Flowchart in recent years in favour of Data Flow Diagrams. However, while acknowledging some problems associated with its construction, it is felt that the System Flowchart's clarity and simplicity make it a most useful tool for describing the flow of operations and control in the existing system. Furthermore some methodologies restrict Data Flow Diagram development to logical analysis and design and so an alternative method of representing the current physical system is required.

5.2 DOCUMENT FLOW: THE SYSTEM FLOWCHART

It was noted in the previous chapter that the analysis of documents is an important task because the existing forms and reports represent the information flow and contents of the current system. A detailed form was introduced to record the structure of documents used in the present system.

However, this standard form gives a relatively static view of the data, failing to clearly show the sequence and interaction of documents. Understanding these latter dynamic aspects of the system is important because any replacement system is likely to affect and change such flows and so the analyst must first be sure that they are properly understood.

A commonly used tool for documenting current flows is the System Flowchart. This summarises what operations are undertaken and where and when they take place.

The System Flowchart is divided into columns with the name of each department, function or personnel written at the top. Inputs from outside the context of the flowchart are shown to the left and outputs to the right. Symbols representing the operations undertaken and the documents used are then placed in the appropriate column. Decisions which lead to different actions being taken can also be shown.

NCC System Flowchart symbols are widely used and recognised and these are shown in Figure 5.1.

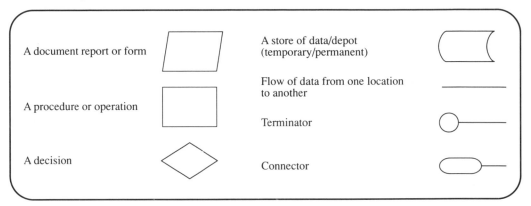

Figure 5.1   Flowchart symbols

The first step in constructing this model is to identify the main departments or functions involved. Appropriate symbols are then drawn in a sequence which gives a general flow of data from top to bottom and left to right. Arrows are used on the connecting lines to indicate the logical flow or sequence where the flow is not in the standard direction. No interaction is implied by crossing lines. The model can then be completed by cross referencing some or all of the symbols to other charts or documents. Cross references can be given in the top stripe of the symbol, whilst the physical characteristics of the input, output or storage may be shown by labelling the bottom stripe.

Figure 5.2 is a System Flowchart for admission procedure in Computer Science at a college. The responsibilities of such people as the course tutor, the admissions tutor and the general office are clearly shown by the position of the symbols. The flow of documents, such as the application form, through the system and the storage of data at different points is also illustrated.

These models are relatively easy to follow and show the sequence of operations in a clear way. They are also fairly intelligible to users and can be employed to support the analyst in an explanation of the current operational systems. The System Flowchart also highlights such issues as excessive information flow, duplication of information stored in different locations, and reasons for delays in particular operations.

A number of guidelines can be offered for drawing System Flowcharts:

— Sort out the column headings before drawing the flowchart. Do not add them as you go along.

— Try to show flows moving from top to bottom and from left to right.

— Attempt to keep flow continuous and self-explanatory. This may demand the addition of explanatory processes which help to illustrate the timing of subsequent actions.

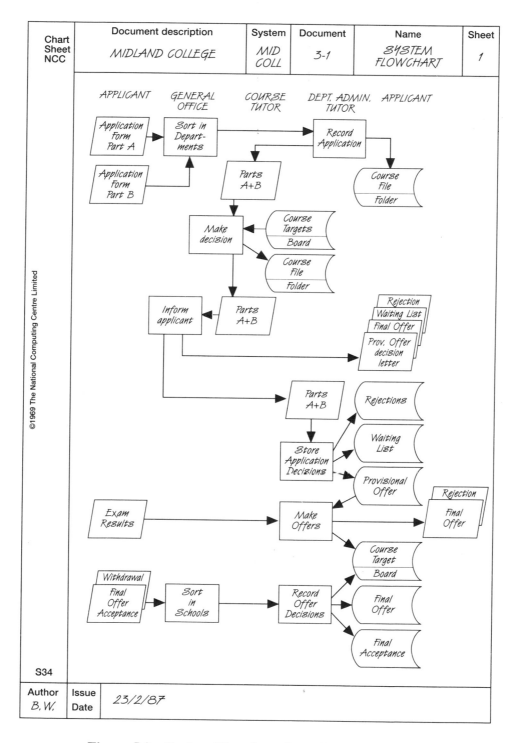

Figure 5.2 System Flowchart for college admissions

— The System Flowchart is a summary model. Connectors can be used to link pages but it is desirable to keep the model relatively small. It is supposed to simplify and clarify – not to just give a visual version of a long narrative.

— There is no 'one best way' of drawing the model. A number of flowcharts, correct in logic but differing in detail, are likely to be equally valid.

— A certain amount of confusion is caused by an overlap between a decision and a process. For example, which of the following is correct?

The answer is that both are acceptable, but that each will be appropriate in different circumstances. The logic of the chart is more important than the shape of the boxes.

— All 'loose ends' should be tidied up. So if there are three parts of an order form each should end up at some clearly defined point – probably in an appropriate permanent file.

5.3   DATA FLOW DIAGRAMS

Data Flow Diagrams (DFD) show the passage of data through the system. They focus on the processes that transform incoming data flows (inputs) into outgoing data flows (outputs). The processes that perform this transformation normally create and use data (data stores). External entities send and receive data flows from the system.

Most users perceive systems from a process perspective and this is reflected in their descriptions of the systems during fact finding.

"We receive these day sheets and amalgamate them into a regional summary".

"The timber is available for dispatch as soon as it has cleared customs".

Consequently a graphical process model such as the Data Flow Diagram is normally very accessible to users and operators of the system. Its focus on transformations closely reflects how they themselves perceive the system.

Data Flow Diagrams were introduced in two seminal books (de Marco, 1979 and Gane and Sarson, 1980) and they have become central to most structured methodologies. In SSADM the DFDs are used in three stages of the systems development process.

• Current Physical DFDs. These record the results of conventional fact finding.

• Current Logical DFDs. The logical information processing of the current system.

• Required Logical DFDs. The logical information processing requirements of the proposed system.

This chapter examines their use in the first of these circumstances. The DFD is used to graphically document an operational interview and will provide the focus of a subsequent discussion with the user where the operational procedures will be confirmed.

5.3.1 DFD Modelling Notation

The Data Flow Diagram depicts the passage of data through a system by using five basic constructs.

Data Flows

A data flow is a route which enables data to travel from one point to another in the diagram. The flow is shown as an arrowed line with the arrowhead showing the direction of flow. The flow is given a simple meaningful descriptive name.

Order Details ⟶

Flows may move from an external entity to a process, from a process to another process, into and out of a store from a process and from a process to an external entity. Flows are not permitted to move directly from an external entity to a store or from a store directly to an external entity. In these instances further processes must be defined to establish the stores (e.g. Create Customer Details) or read them (e.g. Compile Management Reports).

It is generally unacceptable to have a flow moving directly from one external entity to another. However, if it is felt useful to show these flows, and they do not clutter the diagram, they can be shown as dotted lines.

In logical DFDs no two data flows should have the same name. If two flows, separated by a process, appear to be identical then it is likely that the process that separates them is not transforming the data in any way and so should be discarded. It is also useful to adopt this unique naming convention for physical DFDs as it leads to a tighter definition of the current system. The data flows moving in and out of stores do not necessarily require names because the store name may be sufficient to describe them. However in some instances it may be useful to use a name where the flow is especially significant or is not easy to discern from an examination of the diagram.

It may be possible to give a combined name for circumstances where many flows move between the same source and destination.

Processes

Processes are transformations, changing incoming data flows into outgoing data flows. A rectangular box denotes a process. All processes are numbered to permit easy identification. Although the numbering generally follows a left to right convention, there should be no rigorous attempt to show priority or sequence.

The name of the process should describe what happens to the data as it passes through it. A suggested notation is an active verb (extract, compute, verify) followed by an object or object clause. For example:

```
┌─────────┐
│ Verify  │
│ Credit  │
│ Status  │
└─────────┘
```

Other active unambiguous verbs include create, produce, retrieve, store, determine and calculate. The word process should be excluded from any name. Not only is it an ambiguous term but the notation of the box already conveys this information.

In physical Data Flow Diagrams the location of the process may be included in the stripe at the top of the process symbol. No attempt is made to show decision logic within the DFD. There is no equivalent to the diamond-shaped decision box of the System Flowchart.

Data Stores

A store is a repository of data, it may be a card index, a database file or a folder in a filing cabinet. The store may contain permanent data (such as a sales ledger) or temporary accumulations (pending documents, daily movements).

Each store is given a meaningful descriptive name. Data stores may be included more than once to simplify the presentation of the Data Flow Diagram. Such multiple occurrences are shown by an additional vertical line within the store symbol. Each store is also given a reference number prefixed by a letter D to represent a computer data store or M for a clerical data store.

In the example given in Figure 5.3 the arrow from the store is single headed and points towards the process. This is to signify that the process does not alter the contents of the store, it only uses the data available.

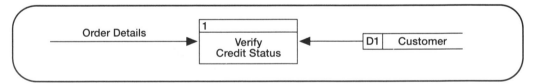

Figure 5.3   Single headed READ arrow

In this way a single-headed arrow shows READ (looking at the data only) or WRITE (changing data only) operations. The creation of the store order is a WRITE-only function of the Verify Credit Status process (*see* Figure 5.4).

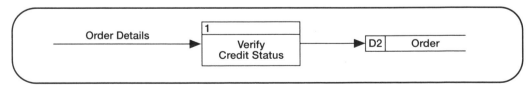

Figure 5.4   Single headed WRITE arrow

In circumstances where the data is both examined and changed (READ and WRITE) then two separate flows are shown. However, to avoid an excessive number of data flows on a high level diagram, double-headed arrows may be used.

External Entities (Source or Sink)

A source or sink is a person or part of an organisation which enters or receives data from the system but is considered to be outside the scope of the project. The source or sink may be duplicated in a completed Data Flow Diagram to simplify presentation. This duplication is again shown by the addition of a line within the symbol.

Because they are considered to be outside the scope of the project, sources and sinks force continual review of the boundaries of the system. As analysis progresses it might be necessary for this boundary to be re-assessed and so certain sources or sinks are brought into the agreed scope of the project. Figure 5.5 demonstrates the notation.

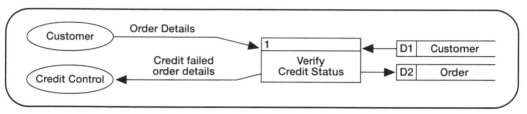

Figure 5.5 Example external entities (Source and Sink)

Physical Resources

A broad arrow is used to show the flow of physical resources and a closed rectangle to represent resource stores. These symbols are not widely used on Data Flow Diagrams because of the clutter produced by their physical size. However, they can be useful for:

— Showing significant resource flows and states.

— Getting started in a project. Users may describe the system in terms of physical flows and stores. The construction of a flow diagram using these symbols permits the identification of the information flows which are needed to support and record the physical movements of goods and materials.

— Finally, and importantly, the physical resources may actually be what the system is about! It is certainly equally important to send both the goods and the invoice. The practical aspects of the system may be lost to an analyst who concentrates too much on the neatness of the data flow.

The permitted connections between the principal components of a Data Flow Diagram are summarised in Figure 5.6.

	Source or Sink	Processes	Stores
Source or Sink	N	Y	N
Processes	Y	Y	Y
Stores	N	Y	N

Figure 5.6   Permitted connections between DFD components

1. The Warehouse Manager receives copies 4,5 and 6 of the order.

2. Copy 4 is placed in a daily delivery folder which contains all orders required for that day. This is used by the Transport Manager to plan his Delivery Schedule.

3. Copies 5 and 6 are given to the warehouse staff to pick the goods from the shelves. Once the order has been assembled it is taken to Despatch where it is placed in a Transit Area awaiting Delivery. Order copies 5 and 6 are given to the Despatch Manager.

4. Goods are loaded and order copies 5 and 6 given to the Van Driver.

5. Both order copies are signed by the customer and Copy 5 returned with Van Driver. Copy 6 remains with the Customer.

6. The Warehouse Manager reconciles the returned Copy 5 with Copy 4. If the data is the same then Copy 4 is marked executed and filed in order number sequence. Copy 5 is passed to Invoicing Section for them to raise the Sales Invoice.

7. If the data on copies 4 and 5 is not the same then the discrepancy must be investigated and an exception report sent to the Managing Director. On successful reconciliation the Copies are amended and forwarded as before.

8. When the goods are picked the warehouse staff must update the 'Bin Card' which sits in a slot in the front of the storage tray.

Figure 5.7   Interview Record: Warehousing system

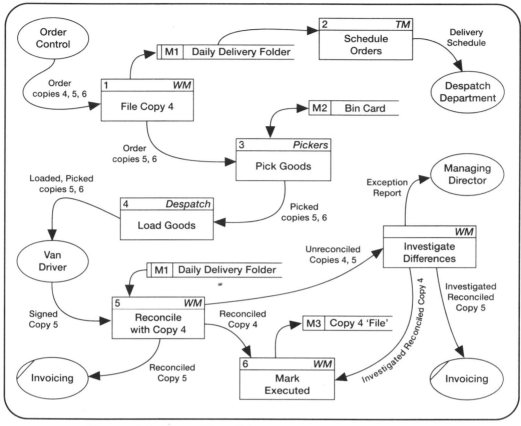

Figure 5.8   Data Flow Diagram: Warehousing system

5.3.2   Modelling Hierarchy

Logical Data Flow Diagrams have a simple and consistent way of representing the successive modelling levels required in a hierarchy. This is discussed in Chapter 7. However, it is also useful to construct physical DFDs at different levels of detail.

In physical Data Flow Diagrams it is likely that the hierarchy will be constructed 'bottom-up'. So, for example, DFDs may be developed for three or four departments of the company and then summarised in an overall diagram. Figures 5.7 and 5.8 are documents produced during the investigation of an order processing system.

Figure 5.7 is the interview record produced after a meeting with the Warehouse Manager of the company. It has been summarised in the Data Flow Diagram given in Figure 5.8. Similar records and DFDs will be developed for Order Processing, Invoicing and Sales which are the other areas under investigation in this particular project.

The DFD shown in Figure 5.8 will be supported by Document Analysis Forms for the following:

— Daily Delivery Schedule.

— Order Form.

— Bin Card.

— Exception Report.

Each DFD will show the activities of each department and will be used to confirm the operations in that particular department. An overall Data Flow Diagram can then be developed by summarising the processes of the departmental DFDs. The selection and naming of these summary processes is largely a matter of choice and experience. However it is usual to establish or choose processes which describe the fundamental activities of the department rather than those describing elements of procedure. So, for example, it is likely that the processes in the warehousing sub-system will appear on the high level order processing DFD (*see* Figure 5.9.) as follows:

— Select Goods: covering processes File Copy 4
                                     Pick Goods

— Confirm Delivery: covering processes Reconcile with Copy 4
                                        Mark executed
                                        Investigate Differences

— Schedule Goods: covering process Schedule Goods

— Load Goods: takes place outside the agreed system boundary.

5.4   ISSUES IN MODELLING CURRENT SYSTEMS

Ed Yourdon (1988) recommended that the systems analyst should avoid modelling the user's current system if at all possible. In coming to this conclusion he makes a number of valid points:

- Some users are dubious about the merits of carefully modelling a system which by definition will be replaced by the development of the new system.

- The current physical models are usually much larger than the logical model because of the redundancy of function and procedure in the current physical system. This is due to the duplication of activity and storage necessary to maintain the integrity of the business operations.

However, he also concedes that the need for physical modelling is based on three major assumptions that turn out to be correct in many projects:

- Drawing the current physical model is a way of educating the analyst in the business procedures of a particular application. The current physical DFD of the warehousing system (*see* Figure 5.8) is a simple way of recording detailed knowledge of a particular area and is also very easy to verify. The user should have little difficulty in confirming this diagram because it contains familiar activities and documents.

- Developing a physical model is a useful way of establishing user confidence. It demonstrates that the analyst understands the business operations and this should provide a firm foundation for subsequent project development. We have already commented on the significance of showing business understanding in interviews

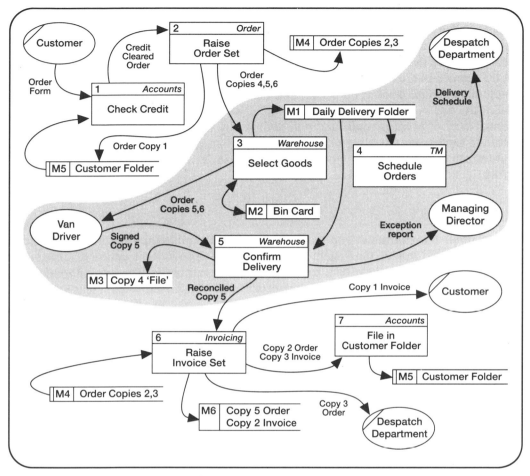

Figure 5.9   Overall Order Processing DFD with warehousing highlighted

(see Chapter 4) and the DFD is a model that can be developed to share and develop that understanding.

- Some users initially find it difficult to work with abstract logical models. They need familiar landmarks to make them feel comfortable, particularly if they have only recently been introduced to the method and products of structured systems analysis.

We would encourage modelling of the current system because of the three reasons given above. Furthermore, the flexibility of manual systems, particularly in dealing with exceptions is very important to recognise and document. Much of the information we gather, particularly about data fields and values, will also be of use to us in the development of logical models.

In many ways Yourdon exaggerates to make a very important point. The danger is that 'the process of developing a model of the current system may require so much time and effort that the user will become frustrated and impatient and ultimately cancel the project'. This is the main point to be made here. Modelling the current system is important but it should not be documented to an extreme level of detail. We have run a case study for many years where timesheets and expense sheets are submitted to a clerk who files them in project folders. The clerk also maintains running cost totals for each project in an exercise book. Delegates on the course are often concerned about when this exercise book is updated – is it on receipt of time or expense sheet? – at the end of the day? – or at the end of the week?. The point is that it does not matter. It is a manual way of holding totals and will completely disappear when the new system is installed (unless we are using software without a totalling facility!). Hence the function is interesting and should be noted but it need not be understood or documented in detail.

5.5 SUMMARY

This chapter has examined two ways of dynamically modelling the current physical system. The first model, the system flowchart, clearly shows the location of stored data as well as highlighting how data passes between departments. The second model, the Data Flow Diagram, concentrates on the processing of data and is less clear about who processes the data and where it is stored. However, unlike the System Flowchart, the DFD will be progressively developed in the following stages of systems development.

Both models can be used to confirm the analyst's understanding of the business system. This confirmation with the user will provide an opportunity for the development of confidence and expertise, with the user being introduced to modelling methods in the context of systems where there are familiar landmarks. However, Yourdon's warnings must be heeded. Too much time spent in this safe territory may lead to the project becoming stuck in the current physical tarpit! This may lead to the user becoming frustrated with the slow progress of the project and lead to its early termination or, perhaps more dangerous in many ways, the rushed completion of the succeeding development stages.

6 Logical Data Perspective

6.1 INTRODUCTION

Data is the raw building block of all information systems and the objective of data models is to express this structure in a concise and useful way. The organisation requires a clear and accurate knowledge of the data structures underlying its information requirements whether they are computerised or not.

Considerable benefits accrue from a detailed analysis of the data structures independent of any hardware and software implementation. For example:

— The data structure tends to be more stable than user`s functional requirements. A business may change over time in how it deals with data and the information required from it, but the underlying structure of this data is unlikely to change as much.

— A lack of understanding of this structure will lead to poorly designed inflexible systems, which may underperform and are costly to maintain.

— Users often find it easier to understand data models which reflect the actual structure of the organisation, rather than the tasks of the enterprise.

This chapter takes a 'top-down' view of the data structures required to support the information systems of the organisation. It concludes with an informal approach to normalisation.

It focuses on the development of the Logical Data Structure (LDS) which shows how the data is logically grouped and related in the context of the business requirements of the system.

6.2 THE MODELLING NOTATION

6.2.1 Entity

This is the term used to describe something that the enterprise recognises in the area under investigation and wishes to collect and store data about. An entity might be:

— Physical – For example; Lecturer, Room, Venue.

— Conceptual – For example; Sales Area, Specialty.

— Active – Course Running, Visit, Lecture.

The forms used in the current system might suggest entities (Order Form suggests Order, 3-part invoice suggests Invoice) but this has to be handled carefully. Most reports will not be entities but summaries of entities, for example, Sales Report suggests the entity Order – not Sale or (even worse) Sale Report. In all instances the entity should be capable of being uniquely identified.

Entities are represented on an LDS by a soft box. Each entity has a meaningful name, normally a noun, which is always singular. The name is placed inside the 'soft box' (*see* Figure 6.1).

BOOK

Figure 6.1   An example entity

It is important to distinguish between the entity type and the entity occurrence. The entity type is Book whilst the entity occurrence is a specific example of Book, such as "Data Analysis for Data Base Design" or "System Analysis". The term entity is used throughout this book as encompassing both of these definitions with its specific use defined by context. The physical equivalent of an entity type is a file and of entity occurrence a record.

Entities are described by attributes (or more accurately attribute types). These are characteristics that represent the entity. For example, the entity Book might be described by the attributes title, author-name, publisher and price. Attribute names are often hyphenated and although this is probably a throwback to COBOL (Benyon,1990) it serves to highlight how the term is being used. Attributes are also called data items or data elements in this book. Its physical equivalent is a field.

A particular entity occurrence should be recognised by the values of an attribute or combination of attributes. For example a Member may be identified by the attribute member-name or a specific Book recognised from the combination of two attributes author and title. This identifying attribute or combination of attributes is termed the entity identifier. Its physical equivalent is a key.

The choice of the entity identifier must be guided by its ability to uniquely identify an entity occurrence. The example given above for Book is unlikely to be acceptable because it is possible that two books will have both titles and authors that coincide. This is not to say that such an identifier will always be unusable. In a small book club with a limited range of stock it is possible that such an identifier will be acceptable. However, in a larger system the analyst will need to introduce an attribute to guarantee uniqueness. All books are given a unique ISB-number and so there is a powerful argument for including this attribute in the entity definition and for making it the entity identifier.

The initial entities will be the objects about which the system currently holds data. These will be identified from the interview notes, documents and observations made in the fact finding and investigation of the current system. Existing file content and searching and retrieval procedures give vital clues to entities of the system.

Figure 6.2 shows the current physical DFD of the order processing system developed in the previous chapter.

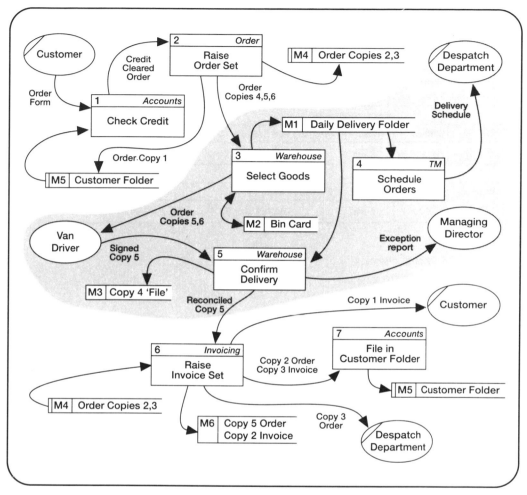

Figure 6.2 Data Flow Diagram of the order processing system

The stores of this system suggest that we wish to hold information about orders, customers, invoices and bin cards. On closer examination of the Document Analysis for bin card it emerges that its identifier is product-code (*see* Figure 6.3). Consequently the entity is named Product which has less physical connotations than bin card. Hence four initial entities are suggested:

Order	identified by	Order-no
Customer	identified by	Customer-code
Invoice	identified by	Invoice-no
Product	identified by	Product-code

6.3 RELATIONSHIPS

A relationship is a relevant business connection between two entities. A relationship is represented on an LDS by a line linking the associated entities.

ADAPT : Document Analysis			System *Example*		Ref: DA *1*	

Document Name *Bin Card*		Ref. No. *—*	Author *SD*	Date *11/03/92*	Page *1* of *1*	

Preparation					Storage			
By	Method	Frequency	Medium	Size	Parts	Order	Location	Duration
Stores staff	*Handwritten*	*1/bin*	*card*	*A5*	*1*	*–*	*Bin*	*–*

Volumes				Use			
Max	Min	Average	Growth	Users		Frequencies	
2/bin	*1/bin*	*1/bin*	*–*	*Stores staff*		*2/week*	

Purpose	Comments
To record physical stock levels	

Ref No	Item	Format	Frequency	Value	Data Source
1	*Prod. code*	*XX999*	*1/card*	*Valid stock no.*	*Prod. Catalogue*
2	*Prod. description*	*Character*	*1/card*	*Valid description*	*"*
3	*Minimum stock*	*99*	*"*	*Numeric +ve*	*Stores Manager*
4	*Movement date*	*Date*	*1/cardline*	*Valid date*	*Clerk*
5	*Intials*	*XXX*	*"*	*–*	*"*
6	*Ref. no.*	*99/999X*	*"*	*Valid order no.*	*"*
7	*In*	*99*	*"*	*–*	*"*
8	*Out*	*99*	*"*	*–*	*"*
9	*Balance*	*999*	*"*	*–*	*Bal=Bal+In-Out*

Figure 6.3   Document Analysis Form for Bin Card

Relationships are often of degree one to many (1:m). For example, an Employee is allocated to exactly one Office, but each Office must have one or more Employees allocated to it. We express this by saying that there is a relationship of one to many between an Office and an Employee. The many aspect of a relationship is represented by the "crow's foot" symbol on the end of the relationship line (Figure 6.4).

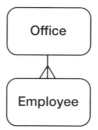

Figure 6.4   Example one to many relationship between two entities

In the simple order processing entity model there appears to be an obvious 1:m relationship between Customer and Order. A customer will place one or many orders but a particular order will be placed by only one Customer. This can be confirmed by a simple occurrence diagram (*see* Figure 6.5).

Returning to the office employee example (Figure 6.4). If each office was allocated to only one employee and each employee allocated to only one office then the relationship between Office and Employee would be one to one.

In certain methodologies 1:1 relationships are not permitted. It is usually suggested that the two entities are merged and that one of the identifiers is selected to identify the merged set.

However, there are dangers in merging two entities involved in 1:1 relationships particularly where the entities do not have the same identifier. In the case of the one-office-to-one-employee relationship, if the entities are merged there is a question of what the new identifier will be. If the identifier is employee number, then when the

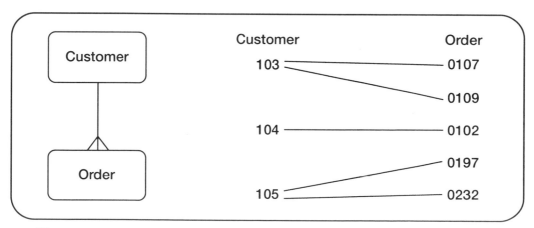

Figure 6.5   Occurrence diagram for the relationship between the entities Customer and Order

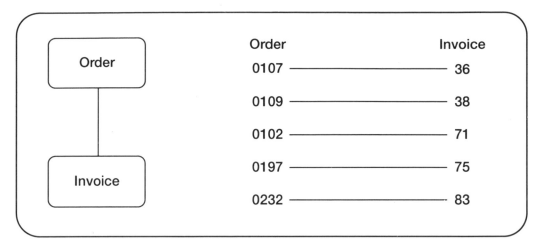

Figure 6.6   Occurrence diagram for the relationship between the entities Invoice and Order

employee leaves the company it is likely that the entity occurrence will be deleted, with all its associated attributes. This means that if the office is not reallocated prior to the occurrence deletion then the attribute values of the office (office number, floor number, square footage, telephone number, etc) will be lost. However if the office number is made the identifier it means that details of new employees cannot be recorded until they have been allocated an office that is free.

In the order processing system the user confirms that there is a 1:1 relationship between Order and Invoice. We agree to model this at present and an occurrence diagram is produced for confirmation (*see* Figure 6.6).

Many to many (m:n) relationships occur quite regularly. For example, an employee may be contracted to work on many projects and a project may have many employees contracted to it (Figure 6.7).

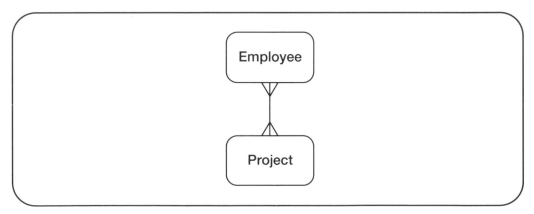

Figure 6.7   Example of a many to many (m:n) relationship

Many to many relationships are normally decomposed into two 1:m relationships with the definition of an additional link entity (Figure 6.8).

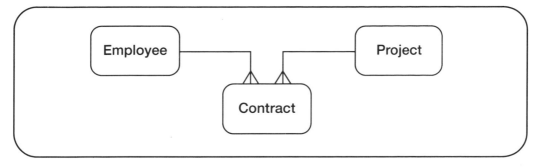

Figure 6.8   Decomposition of a m:n relationship

This allows all the contract details which an individual employee is associated with to be accessed as detailed entity occurrences. It also allows access from the project entity to all the contracts associated with a specific project occurrence. Attributes of the original relationship can now be recorded as attributes of the link entity, for example, date employee contracted to the project, duration of each contract. Note that the name of the link entity is normally the noun form of the verb which described the relationship, thus the contracted relationship is replaced by a Contract entity.

In the order processing example there appears to be a many to many relationship between Product and Order. An order may be for more than one product and we would expect most products to be on more than one order. This is clearly shown in the occurrence diagram (Figure 6.9). This is solved by introducing a link entity (Order Line) which has a one to many relationships with both the original entities (*see* Figure 6.10).

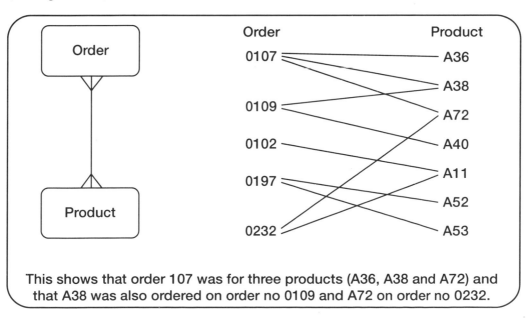

This shows that order 107 was for three products (A36, A38 and A72) and that A38 was also ordered on order no 0109 and A72 on order no 0232.

Figure 6.9   Occurrence diagram showing the relationship between the two entities Order and Product

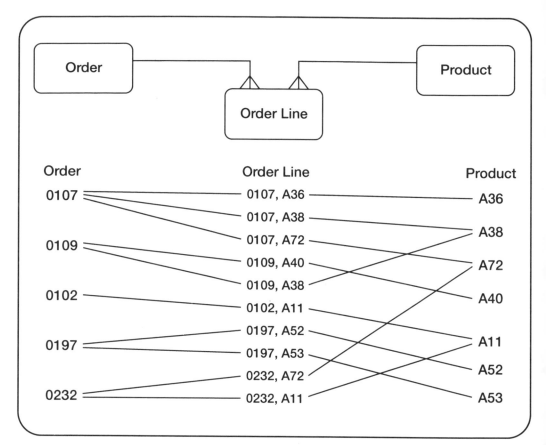

Figure 6.10   Occurrence diagram showing the effect of the new link entity Order Line

Many to many relationships are problematic for at least two important reasons.

- First, they may tend to mask very unsatisfactory parts of a Logical Data Structure. Most many to many relationships often disguise omitted entities. Two examples have already been given above.

- The second reason is both practical and overwhelming. Data Base Management Systems (DBMS) do not support many to many relationships directly. Thus these relationships must be decomposed into two 1:M relationships before they can be mapped on to the software.

However, there are occasions where many to many relationships might be left on the LDS. Some methodologies use these models in the Strategic Planning stages of the project. It might be argued that most managers would be happy with retaining the many to many relationship between Course Run and Delegate (Figure 6.11) as they find it easy to accept that each running of a course would have many delegates on it and each delegate could attend more than one course.

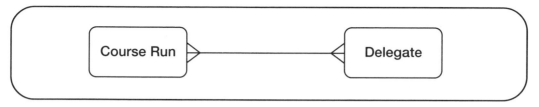

Figure 6.11   Many to many relationship between Course Run and Delegate

This seems more intuitive than the inclusion of a link entity which might be meaningful in terms of data design but less accessible from the point of view of describing the business enterprise (Figure 6.12).

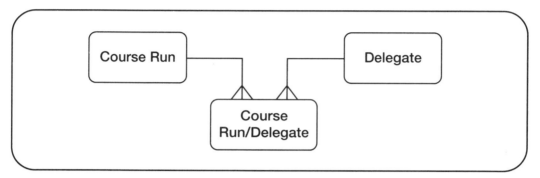

Figure 6.12   Introduction of a link entity between Course Run and Delegate

The full Logical Data Structure for the order processing system is given in Figure 6.13. It is drawn using the 'live crows' approach with the many or detailed entities placed below their logical masters.

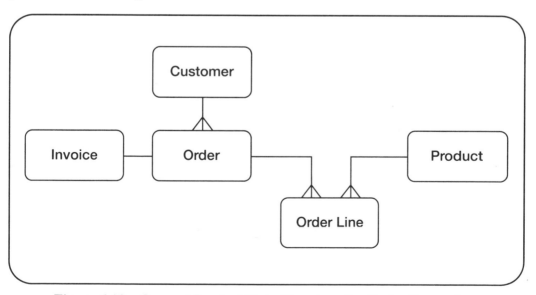

Figure 6.13   Current Logical Data Structure for Order Processing

6.3.1   Relationship naming

The nature of the relationship between two entities is clarified by relationship naming and identification. A relationship link phrase is constructed from the perspective of each entity. For example:

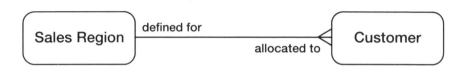

This reads from the Sales Region end as:
  'each Sales Region must be defined for one or more Customers.'
and from the Customer end:
  'each Customer must be allocated to one and only one Sales Region.'

A relationship is mandatory if an entity occurrence cannot exist without taking part in the relationship. For example, each occurrence of Sales Region must contain at least one Customer. If this is not true then the relationship is optional. A dashed line is used to represent an optional relationship. It shows that the entity occurrence at that end of the relationship need not be associated with entity occurrences at the other end.

Relationship naming adheres to the following convention:
each
Entity Name
must be / may be
link phrase
one and only one / one or more
Entity Name

Returning to the order processing example (Figure 6.14). This may be interpreted as;

— each Customer may be source of one or more Orders;

— each Order must be placed by one and only one Customer;

— each Product may be contained in one or more Order Lines;

— each Order Line must be made for one and only one Product;

— each Order must be deemed to contain one or more Order Lines;

— each Order Line must be set up for one and only one Order;

— each Order may be contained in one and only one Invoice;

— each Invoice must be raised for one and only one Order.

The initial LDS is drawn to represent the scope of the current system and validated with users. The word scope is very important here. The LDS is logical and so does not show physical aspects of the current system. Hence entities such as Blue Enquiry Form, Employee Master Listing and Order Form are not permissible. It is their logical equivalents; Enquiry, Employee and Order which will be shown on the LDS. However

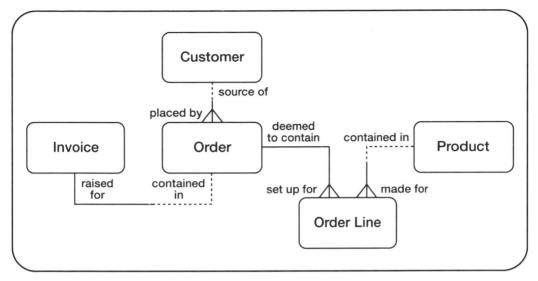

Figure 6.14   Current Logical Data Structure for Order Processing System with named relationships

at this stage the LDS only represents entities identified within current operations and not those associated with requirements. This current LDS can be cross-referenced to the current logical DFD through an Entity/ Data Store grid (*see* Chapter 9).

The LDS is now extended to cover entities and relationships needed to fulfil the new requirements of the system. The extended LDS (Figure 6.15) reflects two requirements.

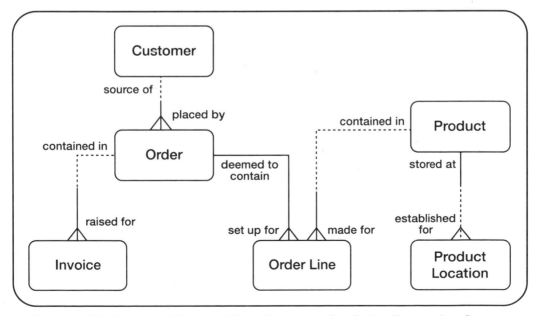

Figure 6.15   Required Logical Data Structure for Order Processing System

- A facility to allow orders to be split across invoices. This now creates a 1:m relationship between Order and Invoice.

- The need to split products across more than one storage location. Storage locations may be empty.

6.4 FURTHER NOTATION AND CONSTRUCTS

A valid LDS can be produced using the basic notation and constructs described above. However the notation and hence information content of a LDS can be enriched by a few simple extensions to this basic notation.

6.4.1   Exclusive relationships

In an exclusive relationship the participation of an entity occurrence in one relationship precludes it from participating in another. For example, in Figure 6.16 an Employee may work from an Office or a Sales Territory but not both. This is indicated by an exclusivity arc.

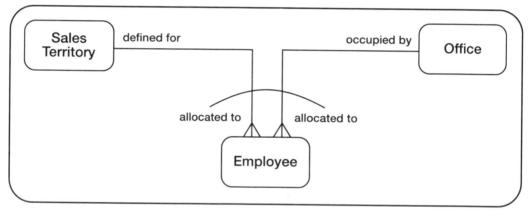

6.16   Example of an exclusive relationship

Example of an exclusive relationship:
- each Employee must be allocated to one and only one Sales Territory OR to one and only one Office;
- each Office must be occupied by one or more Employees;
- each Sales Territory must be defined for one or more Employees.

The exclusive relationship may extend to more than two alternatives.

6.4.2   Recursive relationships

There are situations in which entities have a relationship with themselves. For example, each employee is assigned a supervisor who is also an employee. Each employee who is a supervisor may supervise many other employees. This is an example of a recursive relationship (Figure 6.17).

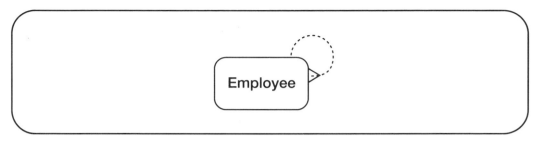

Figure 6.17 An example of a recursive relationship

It must be recognised that data modelling is an iterative process. An LDS is constructed and repeatedly tested against the business requirements of the system.

Validation not only concerns whether the LDS meets the data requirements of the system but also whether it also supports the processing requirements. Each process will need to navigate the data structure by means of the access paths provided by the relationships. As each requirement, and the processing to meet it, is determined the LDS should be examined to ensure that the paths are adequate and will provide the correct processing or answers to queries.

There are no "absolute" entities and attributes. The importance of a particular "thing" will vary with the nature and scope of the system under consideration. For example, a Book may be identified as an entity in the Book Club because it has evident importance to the enterprise. However, in another context, Book may appear as an attribute of an entity as in a Personnel Records system for consultancy staff where it is used to record published works. Thus the context of the model is very important – one system's attribute may be another's entity – because data modelling is not a substitute for the perspective of common business sense.

6.5 ORGANISING THE ATTRIBUTES

We have allocated identifying attributes (identifiers) to entities already. However, there will also be other attributes or data items which we have collected through analysis that we will wish to add to the entity descriptions. Many of these will come from the understanding of current data gleaned through document analysis.

In the order processing example the following attributes have been located

Customer-name	Product-description
Product-price	Bin-location
Amount-stored-in-bin	Order-date
Order-qty	Invoice-date
Customer-address	Customer-tel-no
Customer-type	Customer-credit-limit
Customer-credit-line	

These must now be allocated to appropriate entities. An intuitive allocation might suggest the following (Figure 6.18)

Customer

Customer-code, customer-name, customer-address, customer-type,
            customer-credit-limit, customer-credit-line, customer-tel-no

Product

Product-code, product-description

Product Location

Product-code, Bin-location, amount-stored-in-bin

Order

Order-no, order-date

Invoice

Invoice-no, invoice-date

Order Line

Order-no,Product-code, order-qty,product-price

Figure 6.18   Intuitive allocation of attributes to entities

A consideration of the functional dependency of data items is very helpful in this allocation process.

The formal definition of functional dependency is:

'An attribute Y of an entity R is functionally dependent on another attribute X of R if and only if each value of X is associated with one and only one value of Y.'

For example, department-name is functionally dependent upon employee-code if and only if each value of employee-code is associated with only one value of department-name. However, an employee-code is not functionally dependent on department-name because there will normally be more than one employee working in any one department.

A simple way of showing functional dependencies is to draw a functional dependency diagram as in Figure 6.19.

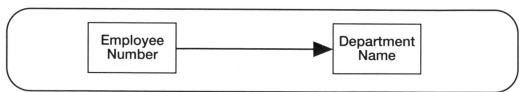

Figure 6.19   Functional Dependency Diagram

Establishing functional dependencies is part of the process of understanding what data means. The fact that department-name is functionally dependent on employee-number comes from the recognition that each employee works in precisely one department. This functional dependency reflects a business rule of the enterprise and must be validated by the users of the system. Many system problems are caused by the developer of the computer system assuming or imposing functional dependencies which do not reflect actual business operations and requirements.

So, for example, if it is necessary to keep a record of all the departments that an employee has worked in since joining the company then department-name would be functionally dependent on employee-code and transfer-date combined. Figure 6.20 shows that department-name is fully functionally dependent on both the attributes employee-code and transfer-date and not on any subset of the group.

A partial, or part key, functional dependency is shown in Figure 6.21 where employee-name is dependent on only a subset of a group of attributes, in this case employee-code.

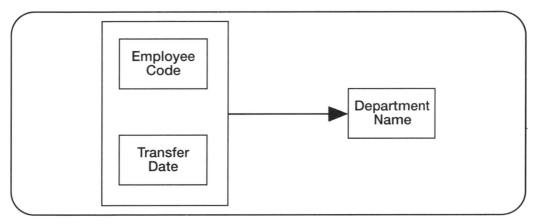

Figure 6.20   Full Functional dependency on a group of attributes

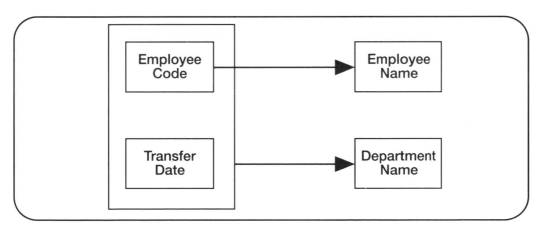

Figure   6.21 Partial functional dependency

It is not necessary to represent all functional dependencies diagrammatically. However it may be desirable to select instances where there appears to be complex inter-relationships between the attributes. The diagrams can also be an economic and unambiguous method of documenting business system rules.

6.5.1   Determinant

A determinant is defined to be any attribute, or group of attributes, on which some other attribute is fully dependent. In Figure 6.21 employee-code is a determinant of employee-name while employee-code and transfer-date together is the determinant of department-name. The concept of determinant is useful in recognising possible identifiers of an entity.

In the allocation of attributes to entity descriptions we wish to ensure that all attributes not in the identifier are fully dependent upon the whole of the identifier of that entity. So, for example, given a value for Customer-code (the identifier of Customer) there is only one value for customer-address, customer-name and customer-tel-no. Where there is more than one attribute in the identifier then it is necessary to ensure that all attributes not in the identifier are dependent upon the whole of the identifier not part of it.

For example, the following description has been developed for the entity Order Line.

Order Line
Order-no, Product-code, order-qty, product-price

Order-qty depends upon both attributes in the identifier because it represents the number of products requested on a particular order. However, product-price depends only upon Product-code. The price does not change with every order that requests it. In this case product-price is re-allocated to an entity in which it is fully dependent on the identifier. If one does not exist then a new entity must be created with the part-identifier becoming the whole of the identifier of the new entity. The decomposition for Order Line is given below.

Order Line   **Order-no, Product-code**, order-qty

Product   **Product-code**, product-price

Furthermore, we would also wish to guard against hidden relationships between data items not in the identifier. For example, consider the entity description developed for Customer.

Customer

Customer-code, Customer-name, Customer-address, Customer-type,
Customer-credit-limit, Customer-credit-line, Customer-tel-no

On further investigation it emerges that customer-type determines customer-credit-limit. In these circumstances customer-type must be examined to see if it is a possible identifier for the whole entity. Obviously in this case it is not (A value of customer-type clearly does not return one value of customer-address). In this circumstance the dependent attribute must be taken out of the entity and inserted in a new entity (which might have to be created) where it is fully dependent upon the identifier. This is shown below.

Customer

Customer-code, Customer-name, Customer-address, Customer-type,
Customer-credit-line, Customer-tel-no

Customer Type

Customer-type, Customer-credit-limit

However, the attribute Customer-tel-no in Customer also seems to create some unwanted data dependencies. A value of Customer-tel-no is likely to be a possible or candidate identifier for the whole data set. If this is so (and in this case it is) then this is acceptable. Customer-code has been selected as an alternative to telephone number because it is shorter and more stable. Customer-tel-no and customer-code are both candidate identifiers for the whole entity but in this case the latter has been selected as the primary identifier.

The discovery of a new entity, Customer Type, raises the issue of how it is related to other entities on the LDS. In fact it is fairly intuitive to realise that Customer Type will have a 1:n relationship with Customer. However, it should also be noted that the identifier of Customer Type is repeated in Customer. In fact relationships are represented through data in this way. The 'many' end of the entity must have the identifier of the '1' end in its own data attribute set. In some cases this identifier also appears in the identifier of the other entity (for example, the identifier of Product, product-code, is also part of the identifier of one of its detail entities (Order Line). In other cases the identifier of the master entity is a non-identifier attribute of the entities

---

Customer

Customer-code, Customer-name, Customer-address, Customer-type,
Customer-credit-line, Customer-tel-no

Customer Type

Customer-type, Customer-credit-limit

Product

Product-code, Product-description, Product-price

Product Location

Product-code, Bin-location, amount-stored-in-bin

Order

Order-no, order-date, customer-code

Invoice

Invoice-no, invoice-date, order-no

Order Line

Order-no,Product-code, order-qty

Figure 6.22   Final allocation of attributes to entities

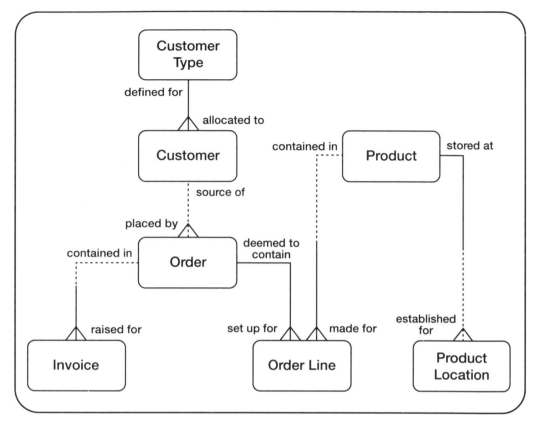

Figure 6.23   Required Logical Data Structure (final draft)

it is related to. For example, the identifier of Customer Type is a non-identifier attribute of Customer. This existence of this so-called posted identifier in the detail entities leads to a further re-consideration of the data sets created in Figure 6.18. The relationships between Customer and Order and Order and Invoice must be reflected in the entity descriptions. This is taken into account in Figure 6.22

The final Required Logical Data Structure is shown in Figure 6.23 and an example entity description is provided for Customer in Figure 6.24.

Hence it is recommended that attributes are added to the entities where their values depend upon the identifier, the whole of the identifier and nothing but the identifier. The aim is to produce a set of entities in which every determinant is a candidate identifier of the whole entity. Entity descriptions arranged in this way are said to be in Boyce-Codd Normal Form (BCNF).

Normalisation is described more rigorously in the complementary text – *Introducing Systems Design*. Normalisation effectively produces a set of data groupings which are known to be free of problems as well as providing a flexible and robust basis for data design. It can be approached in a number of ways. In this text we have adopted a 'top-down' analysis in which attributes are progressively added to the entity descriptions as they are uncovered. This addition is performed under the auspices of BCNF. Benyon (1990) calls this the universal relation approach to normalisation.

SYSTEM	DATE	AUTHOR	No: *17*		PAGE
SOP	*03/02/92*	*SRS*			*1 of 1*

ENTITY DESCRIPTION *CUSTOMER*				

DATA ITEM	TYPE	LEN	DEC	DESCRIPTION
Customer-code	*C*	*5*	*-*	
Customer-name	*C*	*30*	*-*	
Customer-address	*C*	*30x*	*4*	
Customer-type	*C*	*2*	*-*	
Customer-credit-line	*N*	*8*	*2*	
Customer-tel no	*C*	*12*		

Figure 6.24   Entity description for Customer

6.6   SUMMARY

Logical Data Models are found in most structured techniques. Alternative names for this model include entity model, data entity model, entity-relationship diagram and ERA model (Chen 1976, Howe 1983, Benyon 1990, Martin and McClure 1985). It is, with variations, the most standard way of describing data structures in the system and how these structures are related to each other. An alternative modelling notation (Chen) is given in Chapter 10.

This chapter has also introduced a 'top-down' approach to normalisation and used BCNF as a guideline for allocating attributes to entities. A 'bottom-up' approach to normalisation is considered in the complementary text – *Introducing Systems Design*.

The Logical Data Model is developed simultaneously with the Data Flow Model and hence it is important to ensure that the two are consistent. This is formally achieved by cross-referencing the stored data of the DFM (data stores on the DFD) with the stored data of the LDM (entities and their supporting descriptions). The data items of the DFM and the attributes of the LDM are, of course, the same thing. A grid for supporting this cross-reference is presented in Chapter 9.

7 Logical Process Perspective

7.1 INTRODUCTION

This chapter returns to the process model of the system – the Data Flow Diagram. It shows how the current physical DFD can be converted into its logical equivalent – the current logical Data Flow Diagram. This is then amended and extended to reflect the requirements of the replacement system – the required logical Data Flow Diagram.

7.2 CURRENT LOGICAL SYSTEM

Re-casting the current physical Data Flow Diagram into its logical equivalent generally requires four tasks:

- Rationalisation of data stores.
- Rationalisation of processes.
- Renaming data flows.
- Walkthrough checks for consistency and completeness.

This can be illustrated using the current physical DFD for the overall order processing system (*see* Figure 7.1)

7.2.1 Rationalisation of data stores

This is generally achieved by merging data stores which have identical keys or identifiers.

In this example the data stores are as follows

— M1 Daily Delivery Folder   becomes   D1   Order

This is a folder containing the Copy 4 orders. Each order is distinguished by its order number. Hence the logical equivalent is Order with the identifier of order-no.

— M2 Bin Card   becomes   D2   Product

A Bin Card exists for each product held in stock. It records how much is in the 'bin' or stock location. The analyst would have to investigate whether a bin could contain many products or whether, and more likely, a product could be stored at more than one

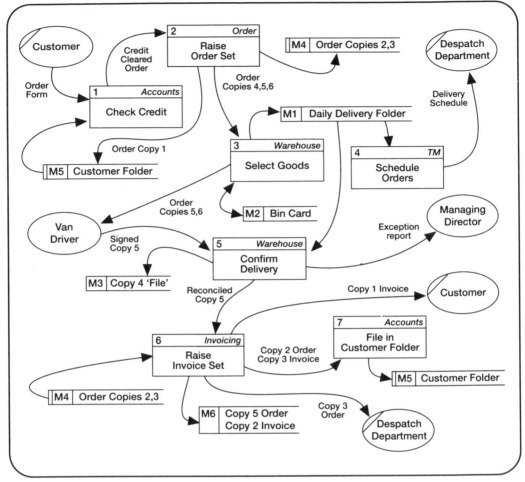

Figure 7.1   Overall Physical DFD – Order Processing

location. In this example it is assumed for now that a product is only stored in one location and that this location only contains one product type. Hence each bin card can be uniquely identified by product-code which is the identifier of the logical store Product.

— M3   Copy 4 'File'   becomes   D1   Order

This file is simply a later state of M1.

— M4   Order Copies 2,3   becomes   D1   Order

The fact that there are physically two copies of the order in this store is irrelevant in the Logical Data Flow Diagram.

— M5   Customer Folder   becomes   D3   Customer

A folder is created for each customer into which Copy 1 of the order is placed upon order creation (process 1) and Copy 2 order and Invoice Copy 3 after invoice creation (process 7). The identifier of each record set (folder) is customer-no and the logical store is best described as Customer.

— M6	Copy 5 order	becomes	D1	Order
	Copy 2 invoice	becomes	D5	Invoice

The relationship between order and invoice needs clarifying. However, the Logical Data Model (*see* previous chapter) has already established that these are separate entities. If we subsequently find that there is a 1:1 relationship between order and invoice then we may merge these two logical stores and only retain D1 – Order.

Thus the current logical DFD will show the logical stores:

D1 Order
D2 Product
D3 Customer
D4 Invoice

It is usual for logical DFDs to have considerably fewer stores than their physical equivalents.

7.2.2 Rationalisation of processes

General rules about the rationalisation and grouping of processes in Data Flow Diagrams are difficult to give. A number of possibilities exist and will be applicable in different circumstances. For example:

— Remove 'reorganising' processes such as Sort Locations or Index Applications.
— Remove 'retrieve only' processes and record in the Requirements Catalogue.
— Remove processes which will be dealt with outside the system.
— Combine processes which are essentially a series of activities.

Each application must be dealt with on its merits. These may be assessed in walkthroughs which check for consistency, completeness and clarity.

If the current physical DFD has been drawn at an appropriate level then it is likely that most processes will progress unchanged to the logical stage. However, the current terminology needs examining to see if there are any undue references to the physical nature of the activity. Slight changes may result. For this example:

Old process name		New process name	
1	Check Credit	1	Check Credit
2	Raise Order Set	2	Create Order
3	Select Goods	3	Select Goods
4	Schedule Orders	4	Schedule Orders
5	Confirm Delivery	5	Confirm Delivery
6	Raise Invoice Set	6	Create Invoice
7	File in Customer Folder	7	——

Process 7 is purely procedural. It logically updates the Customer store and this is now integrated into process 6 to preserve the logic of the operations of the system. This is an example of combining processes which are a series of activities. On a Logical DFD

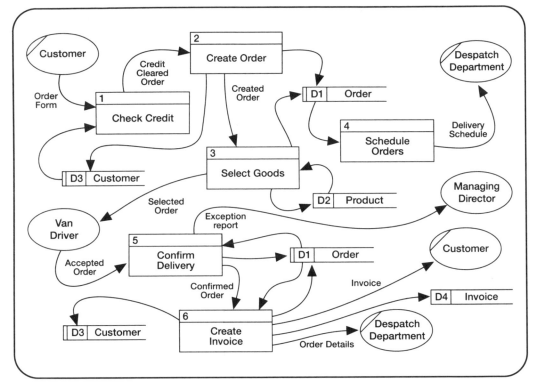

Figure 7.2   Current Logical DFD for the Order Processing System

all physical locations are removed from the top stripe of the process box.

7.2.3   Renaming of data flows

Data flows should be examined and references to physical media, copies and colours removed. It must be recalled that each data flow should have a unique name on the logical DFD.

The results of logicalisation are shown in the current logical DFD of Figure 7.2.

7.3   MODELLING HIERARCHY

The Data Flow Diagram has a simple and consistent way of representing the successive modelling levels required in a hierarchy. Each process is exploded into a lower level Data Flow Diagram until the process can be represented in an Elementary Process Description. In this way it is possible to present a series of Data Flow Diagrams representing increasing levels of detail and appropriate for different types of task and staff. The convention of denoting the highest level diagram as level 1 has been adopted with subsequent levels designated 2, 3, etc. Indeed the concept can also be extended backwards where the complete level 0 Data Flow Diagram is a one process diagram which summarises the inputs and outputs of the system under consideration. This is called the context diagram.

The numbering system of the Level 1 model needs to be extended to the lower level diagrams so that they can easily be referenced back to their 'parent' processes. This is achieved by using a decimal numbering system. Thus process 3 may be decomposed into 3.1, 3.2, 3.3, etc and, if a further level is required, 3.1 into 3.1.1, 3.1.2, 3.1.3, etc.

It is important to check that all flows and stores accessed by the process in the higher level box are actually used in the lower level decomposition. Data Flow Diagrams should be balanced so that data flows in and out of a process must appear on the Data Flow Diagram that is the decomposition of that process. If a flow does not appear then the reason for its existence at the higher level must be closely examined. Similarly, flows should not be produced that do not exist in the parent process. If such a flow seems to be necessary then all higher level diagrams should be altered so that the whole model remains balanced.

Three general guidelines can be given for controlling decomposition of Data Flow Diagrams.

Decomposition is complete when:

- Each process can be described in an A4 Elementary Process Description.

AND

- There is no process on the lowest level DFD with a read/write flow to a store (because these are usually undertaken by two separate processes).

AND

- An exploratory interface line (showing the human – computer boundary) does not intersect any processes on the lowest level DFD.

Most systems can be modelled within two or three levels. Processes decomposed beyond this often become trivial and do not access stores in any way.

The lowest level of decomposition of any process is shown by an asterisk on the bottom right hand corner of the box. Obviously such processes can occur on any of the levels of the diagram. It is quite feasible and acceptable for processes on the level 1 DFD to not require further decomposition.

7.4   REQUIRED LOGICAL SYSTEM

The Required Logical Data Flow Diagram will reflect new user requirements and solve current system problems. New processes will be defined to reflect new needs whilst activities which are no longer required or are dealt with elsewhere in the system will be discarded.

Problems and Requirements are formally documented in the Problems/ Requirements List (PRL). This was introduced in Chapter 4 and is repeated as Figure 7.3.

Some of the problems of the current system are due to data duplication and so a thorough review of inter-process flows is helpful. Inter-process flows imply transaction at a time processing while batched processes read stored data.

We need to ensure that the logic of the system is correct and that remaining inter-process flows are true reflections of the logical requirements and not the remains of an

ADAPT : Problem/Requirement List			System *Example*	Ref: PR *01*

Author *S.D.*	Date *17/01/92*	User *Agent*	Priority *1*	Page *1* of *74*

Problem/Requirement Description

Currently impossible to distinguish between two (or more) shipments of the same sales or purchase order.

DFD/Entity Ref *17*	Document Ref *Purchase Order*	Other Ref

Solution Description

DFD/Entity Ref	Document Ref	Other Ref

Figure 7.3   Example Problems/Requirements List (PRL)

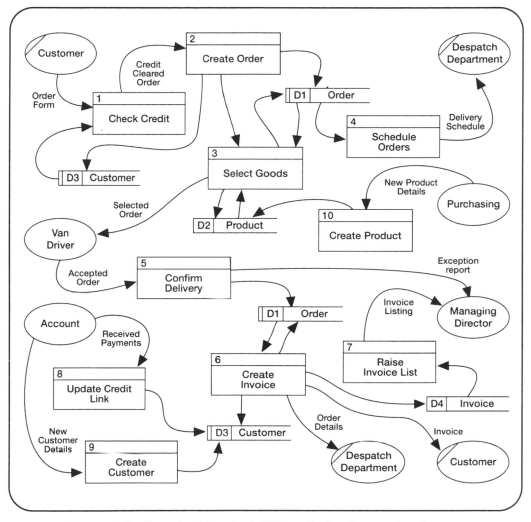

Figure 7.4   Required Logical DFD – Order Processing System

inter-departmental document flow. In the Current Logical DFD for the order processing system the following clarifications are agreed with the user.

- Orders are individually credit cleared
- Selection of goods is performed on a batch of orders
- Invoices are raised on a batch of orders.

These changes are reflected in the Required Logical DFD. Only one inter-process flow remains and this is typical of business administration systems where much of the activity takes place on a group of transactions.

It is also useful at this time to ensure that all data stores contained in the DFD are created by an appropriate process. Examination of the Current Logical DFD shows that the stores Order and Invoice are created but Product and Customer are not. This is investigated and appropriate processes added to the Required Logical DFD (processes

9 and 10).

Requirements will also be formally listed and prioritised in the PRL. The Required Logical DFD reflects all or selected requirements. This usually requires the definition of extra processes and perhaps new stores. Two examples are included in Figure 7.4

R1: To produce an invoice listing for the Managing Director

This is achieved by new process 7.

R2: To update the credit line value on Customer by receiving payment information from accounts.

The new process 8 achieves this.

The Required Logical DFD now fulfils the requirements and solves the problems

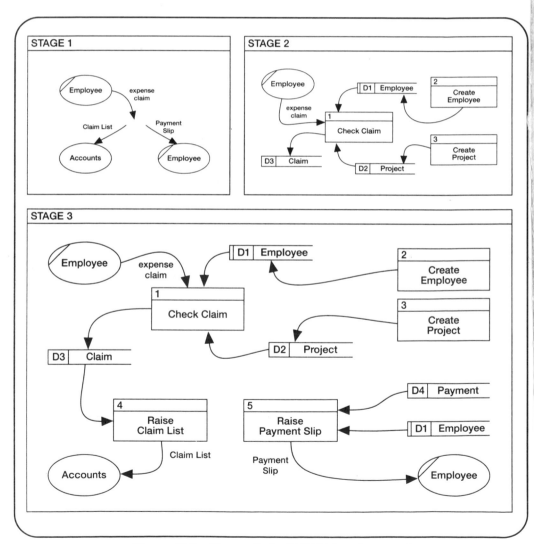

Figure 7.5   Stages 1–3 of DFD development

ı the PRL. It is a logical definition of the system needed to fulfil the user's requirements.

CONSTRUCTING THE MODEL – FURTHER GUIDANCE

e previous sections have described the incremental development of a simple Data ⸱low Diagram. However, it is also possible to give a set of simple guidelines which are particularly helpful if there is no current DFD to build upon. Stages 1 – 4 are demonstrated in Figure 7.5 and 7.6

Stage 1:   Establish the major inputs and outputs of the system

Establish the source of each input and the receiver of each output and make these external entities.

Stage 2:   Establish a process that handles each input on its arrival into the system.

Establish the data stores that this process needs to create or use if it is to be successful.

Stage 3:   Establish a process that generates each output flow from the system.

Establish data stores required by these processes

Stage 4:   Examine the process(es) and associated flows that link the input and output processes.

Establish the rules for these processes and expand the Data Flow Diagram

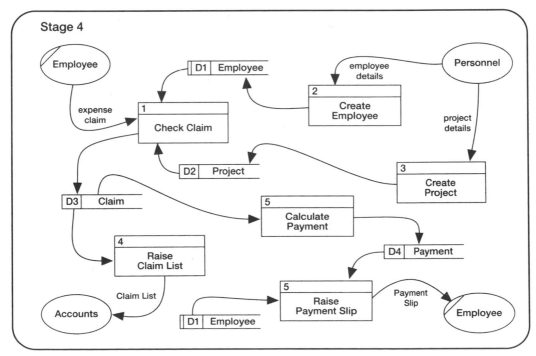

Figure 7.6   Stage 4 of DFD development

to reflect these rules.

Identify other inputs and outputs required by processes.

Stage 5:   Establish a hierarchical set of Data Flow Diagrams which reduces
number of processes in the top level diagram to a manageable number (say
7 to 10).

The simple example does not require a hierarchy but we might simplify an extended
version of the order processing system DFD. Figure 7.7 shows this extended version
which has also introduced procedures for dealing with uncleared orders (process 1).
Both process 1 and 7 are now exploded in detail in lower level DFDs. The appropriate
DFD for process 1 is shown in Figure 7.8.

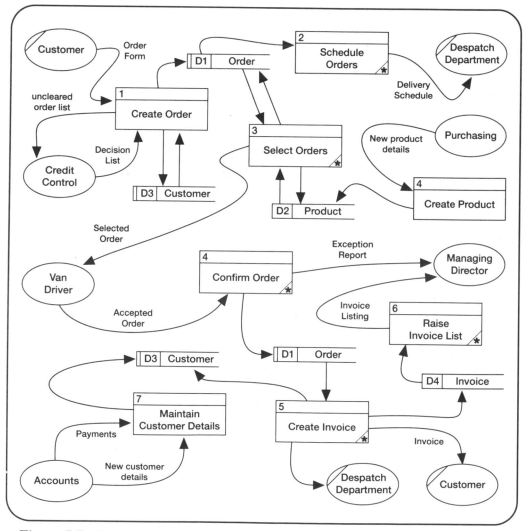

Figure 7.7   Extended and reorganised DFD for the Order Processing System

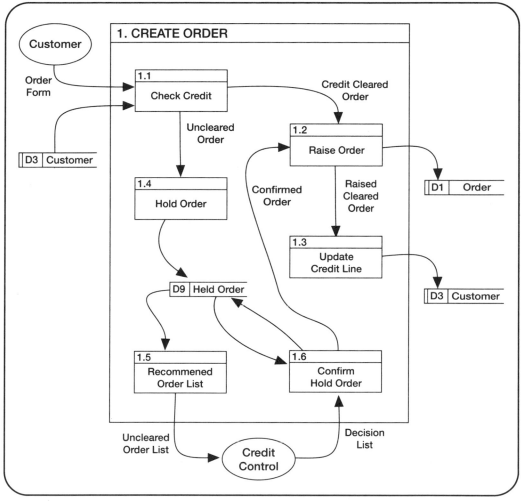

Figure 7.8   Level-2 DFD for process 1 of the Order Processing System

7.6   ELEMENTARY PROCESS DESCRIPTION

The guidelines for modelling hierarchy (given in Section 7.3) suggested that decompostion was complete when the process could be described in an A4 Elementary Process Description. It is recommended that formal description methods are used, not ambiguous English narrative. Three process description techniques are described in this section – decision tables, decision trees and Structured English. A development of the latter, the Action Diagram, is examined in the complementary text *Introducing Systems Design*.

7.6.1   Decision Tables

The following narrative summarises an interview with an accounts clerk in a Mail Order Book Club.

"Priority treatment is given to members who order more than £100 value of books in a year and have a good payment record, or who have been members for over 10 years."

The statement is ambiguous as it is not clear if poor paying members of over ten years standing will receive priority treatment. Drawing a decision table will highlight this ambiguity and the Analyst will be able to precisely express the logic. The following narrative is now more explicit but is hardly conversational!

'Members of the Book Club whose order value exceeds £100 in a year and have a good payment record are given priority treatment. Where a membership has been over 10 years the member will get priority treatment if the order value exceeds £100 per annum, in which case the club is prepared to overlook a poor payment record. If the order placed by a member of more than 10 years standing does not exceed £100 during the year, but the member has a good payment record, then priority treatment is also given. In all other circumstances the order is given normal treatment.'

A decision table is useful for representing the conditional logic of processes where different actions are taken depending upon the occurrence of a particular combination of circumstances. This represents typical discussions where staff explain that " when x and y happen then we will do this and that, on the other hand y might not happen and, in this case, we will do something else...". In such situations the analyst will often feel confused and unsure that the rules of the enterprise have been completely understood. Expressing the logic in the form of a decision table should clarify the rules governing the options.

Drawing a Decision Table

A decision table is divided into the four parts shown in Figure 7.9. The conditions, that determine which actions will result, are listed in the condition stub. Combinations of these conditions are then identified and expressed as rules or condition entries. The possible actions which can occur as a result of different condition combinations are listed in the action stub. The decision table is completed by entering the relevant actions for each condition rule identified.

The simplest type of decision table is a limited entry decision table in which conditions are expressed as questions which may be answered by a simple Yes (Y) or No (N). The condition entries or rules are then specified as combinations of these answers. This type of table is self-checking to the extent that there is one rule for each possible combination of conditions.

			Rules
IF	these conditions apply as specified in the entries	CONDITION STUB	CONDITION ENTRIES
THEN	these actions are to be taken as specified in the action entries	ACTION STUB	ACTION ENTRIES

Figure 7.9   Structure of a Decision Table

The relevant action for each combination of conditions is recorded by an X in the action entries section.

The construction of a Limited Entry Decision Table can be illustrated from the example Book Club narrative.

1. Identify all conditions, being careful not to include mutually exclusive conditions such as Good Payment record and Bad Payment Record. Write the conditions down in the Condition Stub with the the most significant or critical one first. If more than four conditions apply then divide the table in two.

    c1: Good payment record?
    c2: Order value > £100 p.a.?
    c3: Member > 10 years?

2. Identify all actions possible and record them in the Action Stub in the sequence in which they occur.

    a1: priority treatment
    a2: normal treatment

3. In the case of a limited entry table the number of rules can be calculated by using the formula (2 **c). Ensure that all the condition entries or rules have been expressed. In this example, three conditions generates 2**3 rules, since each condition entry can be Y or N. This explains why the division of a table at four conditions was suggested above. 2**5 gives 32 rules, which may be a little unwieldy. A suggested format for allocating the Ys and Ns is given below. For z rules, the first row has z/2 Ys and z/2 Ns, reducing down until the last row which is always YNYNYN, etc.

c1: Good payment record?	Y	Y	Y	Y	N	N	N	N
c2: Order value > £100 p.a.?	Y	Y	N	N	Y	Y	N	N
c3: Member > 10 years?	Y	N	Y	N	Y	N	Y	N
a1: priority treatment								
a2: normal treatment								

4. Action Entries are now made. These are derived by applying each combination of conditions to the actions described in the narrative. The following table is produced.

c2: Good payment record?	Y	Y	Y	Y	N	N	N	N
c1: Order value > £100 p.a.?	Y	Y	N	N	Y	Y	N	N
c3: Member > 10 years?	Y	N	Y	N	Y	N	Y	N
a1: priority treatment	X	X	X		X			
a2: normal treatment				X		X	X	X

5. In limited entry tables there should be 2 ** C rules in total, where C is the number of conditions expressed in the Condition Stub. Check the table completeness by counting the rules.

Table 1

| Places available on course? | Y | Y | N |
Financial support available?	Y	N	–
Go to Table 2	X		
Go to Table 3		X	
Reject application			X

Table 2

| Qualifications satisfactory? | Y | Y | N | N |
Reference satisfactory?	Y	N	Y	N
Make offer	X			
Reject application				X
Go to Table 3		X	X	

Table 3

| Interview satisfactory? | N | Y | Y |
Mature student?	–	Y	N
Offer supported place		X	
Reject application	X		
Place on waiting list			X

Figure 7.10   Hierarchy of decision tables

6. Check the table for redundancy. This is present if two or more rules exist with different combinations of conditions leading to the same actions. In such circumstances consolidate the rules where possible. A consolidation is shown in the Condition Entries by a dashed line. This effectively indicates that this condition is irrelevant – it can be Yes or No – the action remains the same. Two consolidations have been undertaken to produce the following table.

c2: Good payment record?	Y	Y	Y	N	N	N
c1: Order value > £100 p.a.?	Y	N	N	Y	Y	N
c3: Member > 10 years?	–	Y	N	Y	N	–
a1: priority treatment	X	X		X		
a2: normal treatment			X		X	X

7. It is possible to place an ELSE statement at the end of the chart to permit simplification and conciseness.

c1: Good payment record?	Y	Y	N	E
c2: Order value > £100 p.a.?	Y	N	Y	L
c3: Member > 10 years?	–	Y	Y	S
				E
a1: priority treatment	X	X	X	
a2: normal treatment				X

8. Use a 'top-down' hierarchy of tables where more than four conditions are involved. This is useful in circumstances where there are too many conditions operating to result in a simple table. The most important conditions are evaluated in a high level table and depending upon the responses to these the flow is directed to one of several other decision tables. An illustrative example is given in Figure 7.10

Extended entry Decision Tables

The concise nature of the Decision Table has prompted its extension to include circumstances where the Condition Entries are expressed as values of some kind. An extended entry Decision Table may also include symbols or codes in the action entry section of the table.

Consider the following Limited Entry Decision Table. (Figure 7.11)

Member < 1 year	Y	N	N	N
Member 1–5 years	–	Y	N	N
Member 6–10 years	–	–	Y	N
No discount	X			
Discount 10%		X		
Discount 25%			X	
Discount 50%				X

Figure 7.11   Limited Entry Decision Table

Using a Limited Entry Decision Table for this example appears unsatisfactory since the conditions are all related to each other. Only one action can take place and the table looks unnecessarily complex. Figure 7.12 describes the same logical procedure in a simpler way by using an extended entry decision table.

Length of membership	<1	1–5	6–10	>10 years
Discount %	0	10	25	50

Figure 7.12   An Extended Entry Decision Table

Hybrid decision tables can also be constructed where the conditions are a mixture of Yes and No and other values as in Figure 7.13.

Good payment record	Y	N	Y	N	etc........
Length of membership	<1	<1	1<5	1<5	
Accept	X		X		
Reject		X		X	etc.......
Discount %	0		10		

Figure 7.13   Hybrid Decision Table

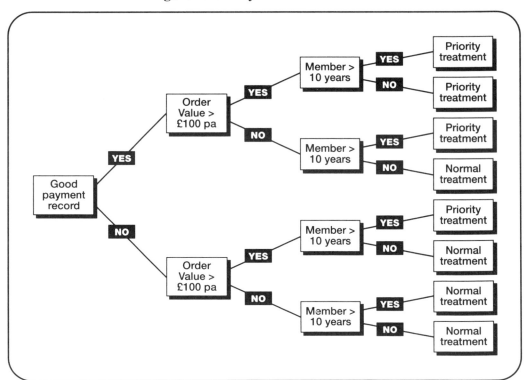

Figure 7.14   Decision Tree

7.6.2   Decision Trees

Decision trees are another way of showing the alternative actions that can result from different combinations of circumstances. The diagram resembles a fallen tree, with a root on the left hand side and branches representing each decision. The tree is read from left to right and the actions to be undertaken are recorded down the right hand side of the diagram. The conditional logic of the Book Club order handling process described in the previous section has been recorded in the form of a Decision Tree in Figure 7.14.

7.6.3   Structured English

Ordinary English narratives often lead to misunderstandings due to the ambiguity of the language. The narrative can be sharpened by using Structured English. This uses standard narrative constructs that constrain the language but, at the same time, permit a certain degree of user understanding. The structure is enhanced by indentation.

```
Book Orders
     Get Member Order details
          IF Good Payment Record
               IF Order Value > £100 pa
                    DO Priority Order
               ELSE (Order Value < £100 pa)
                    IF Member > 10 years
                         DO Priority Order
                    ELSE (Member < 10 years)
                         DO Normal Order
                    ENDIF
               ENDIF
          ELSE (Poor Payment Record)
               IF Order Value > £100 pa
                    IF Member > 10 years
                         DO Priority Order
                    ELSE (Member < 10 years)
                         DO Normal Order
                    ENDIF
               ELSE (Order Value < £100 pa)
                    DO Normal Order
               ENDIF
          ENDIF
```

Processes may be described in Structured English by several basic constructs.

- Sequences of events or actions are described by simply using the top to bottom order of the statements.

  For example.

— Accept application.

— Store applicant data.

— Add to list of applicants.

— Send letter of acknowledgement.

The four actions above are undertaken in the order presented in this sequential structure. The statements are presented in line and all actions are performed.

- Circumstances where actions depend upon differing conditions can be described using Conditional statements such as IF..... ELSE........ENDIF.

- Where one of several possible cases apply statements such as:

<pre>
        CASE 1 ....
        CASE 2 ....
</pre>

Repetition of action can be described using statements such as:

<pre>
        DO WHILE ....
or   DO UNTIL ....
or   FOR ALL.....
</pre>

For example

```
DO WHILE still forms on pile
        Read application details
        IF application is for overseas student
        Request proof of financial support
        ENDIF
        IF course has vacancies
                CASE (exam results meet requirements)
                        Make final offer to applicant
                CASE (results not known)
                        Make a provisional offer
                CASE (results not good enough)
                        Reject application
                ENDCASE
        ELSE (course full)
        ENDIF
ENDDO
```

This clearly shows the hierarchy of constructs (shown in capitals) and comments have been added to clarify the narrative.

Decision Tables, Decision Trees and Structured English are essentially tools for logic or process description. In a comparison that also distinguishes Structured and Tight English (a distinction we have chosen not to pursue) Gane and Sarson (Gane and Sarson, 1980) assess these tools against a set of criteria. They conclude that Decision Tables are very good from the perspective of logic verification and computability but that they are not easy to use or amend. In contrast, Decision Trees are simple, good at displaying the logical structure, but poor from the viewpoint of verification and computability. They arrive at the following conclusions:

— Decision Trees are best used for logic verification or moderately complex decisions which result in up to 10-15 actions. They are also useful for presenting the logic of a Decision Table to users.

— Decision Tables are best used for problems involving complex combinations of up to 5 or 6 decisions. Large numbers of combinations of conditions can make Decision Tables unwieldy.

— Structured English is best used wherever the problem involves combining sequences of actions with decisions and loops.

(Gane and Sarson, 1980)

The computability of Decision Tables makes them a very attractive candidate for automation. One commentator (Lew, 1984) has claimed that program code derived automatically from Decision Tables can be proven correct and hence does not need recasting into conventional languages or flowcharts. A much earlier advocate of the Decision Table (Fergus, 1969) gave examples of four pre-processors, including a 1967 product producing PL/1 source statements. Decision Tables also had a central role in Grindley's SYSTEMATICS specification language (Grindley, 1968).

7.7 SUMMARY

The current and required logical Data Flow Diagrams retain the same notation as the current physical DFD. Hence the notation introduced within a familiar user framework (the current physical system) should be assimilated by the time logical analysis and design commences. Consequently, new ideas and terms are introduced within a consistent notation.

Data Flow Diagrams (DFDs) are a fundamental modelling technique of most structured methodologies. They were introduced by de Marco (1979) and Gane and Sarson (1980) in the two seminal books that laid the foundation of the structured systems movement. They have also been extended to real-time development (Ward and Mellor, 1985 – *see* Chapter 10) and are found in most proprietary methodologies.

A number of process models are available to unambiguously define the Elementary Process Description. Three have been presented in this chapter – Decision Tables, Decision Trees and Structured English. Another candidate, the Action Diagram, is described in detail in Introducing Systems Design.

8 The Logical Event Perspective

8.1 INTRODUCTION

We have so far produced two perspectives of the system:

- The processing view, represented by the Data Flow Diagram, describing the way that data is transformed and stored as it passes through the system.

- The data view of the Logical Data Structure modelling the information structures of the system.

Entity Life Histories (ELH) provide a third perspective of the system requirements. Their development requires consideration of the events which trigger the processes which, in turn, have an effect on the information structures. They provide a mechanism by which the data and functional views can be validated against each other to provide a sound basis for design.

The Entity Life History is a major analysis technique within SSADM. It shows the sequence in which events affect each entity and models time and business constraints. It validates the data views of the system and identifies errors and omissions which lead to further detailed processing and data requirements so producing a complete and consistent specification.

8.2 EVENTS AND EFFECTS

An ELH shows the events which may have an effect on a particular entity occurrence. An event can be defined as something that triggers a process to update system data. An event is not a process, it is the stimulus which causes that process to be invoked. So, for example, the process might be Create Order but the event is Receipt of Order or Order Received. In practice, due to the diagram space constraints, event names are frequently shortened; Examination Result, Examination Notification, Resignation, etc. However, care should be taken that the name reflects the event rather than the DFD process, otherwise there is a danger that other events triggering the same process may be missed.

Three types of event can be distinguished:

- *External event* – A transaction arriving from the outside world. For example, Receipt of Order. These events are normally associated with data flows of the Data Flow Diagram.

- *Internal process event* – This occurs when a predefined condition within the stored data has been met. For example, the Ship Cleared event takes place when all the timber has sold from the timber berth.

- *Time-based event* – This occurs when a particular process is to be triggered at a regular time interval: a set time of day, month, year, etc. For example, the event Pay Wages takes place on each Friday at 1400 hrs.

A single event instance may cause more than one entity occurrence to change. The changes within a single entity occurrence caused by an event is called an effect.

An effect can lead to the:

— Creation of a new entity occurrence:
For example, Receipt of Order causes the creation of an entity occurrence of Order.

— Deletion of an existing entity occurrence:
For example, Resignation may cause the deletion of an entity occurrence of Member.

— Modification of existing entity occurrence:

For example, Payment Receipt causes the modification of the data item Amount-due in the entity occurrence Customer.

Modification is also said to take place when data items in a created entity occurrence are inserted for the first time. For example, the dispatch-date of a particular order is unlikely to be known on the creation of order. It will only be inserted later in the order cycle and is thus treated as a modification of the entity occurrence.

Effects on relationships are also treated as modifications. So, for example, the New Owner Notification event is a modification of the entity occurrence of Car not a deletion followed by a creation.

It is valid for one event to affect an entity occurrence at more than one point within its life or, at different times, to invoke substantially different processing.

There may be occasions when a single instance of an event may affect an entity in one of several mutually exclusive ways. For example, the event Transaction acting on the entity Account might have different effects when the transaction is a payment to when the transaction is a withdrawal. For example, a withdrawal transaction might require a check on the balance of the account and a count of the number of withdrawals within a specified time period. In contrast a payment transaction may only update the account balance and record transaction details. In these circumstances the event name should be qualified by a description of the exclusivity. For example, Transaction(payment) and Transaction(withdrawal). The processing required will be determined at the time of the event either by the stored data or by the data accompanying the event.

If a single event affects more than one occurrence of a particular entity and the effects are different for each entity occurrence then the entity is deemed to be assuming different roles. As processing will need to be specified for each role, each different role must be separately identified on the ELH for the relevant entity. On the ELH the event name must be qualified by the role which the entity is assuming. For example, if the business rules are such that a patient can be registered with only one doctor at any one time, notification of doctor registration would have the effect that the old registration must be cancelled and a new registration created. In this case there are two roles for the one event, Notification of Registration[old] which cancels the current registration, and Notification of Registration[new] which creates a new registration occurrence.

The role names are used to distinguish between the different effects of the event. Note that different brackets are used for role qualification and for mutually exclusive effects.

8.3   CONSTRUCTION OF ENTITY LIFE HISTORIES

The ELH is used to investigate the lives of each entity, identifying the events which have an effect on their lives, documenting the way in which the lives are affected and showing the sequence in which the effects take place. One ELH is drawn for each entity.

Thus an ELH represents all the permitted sequences of events that may occur during the life of an occurrence of an entity. Each occurrence is constrained to behave in a way defined by the ELH for that entity. However, at any one time each entity occurrence may be at a different stage in its life history, and each entity occurrence may lead permitted lives that are different from those of other occurrences of the same type.

The construction of Entity Life Histories begins with the development of the Event/Entity Matrix.

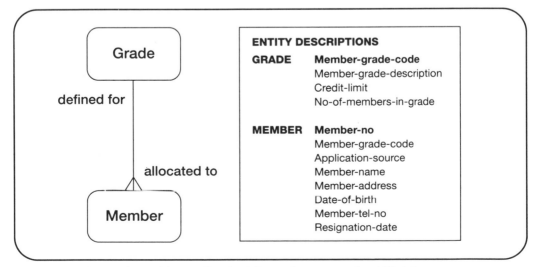

Figure 8.1   Simple Logical Data Structure for 'Club' system

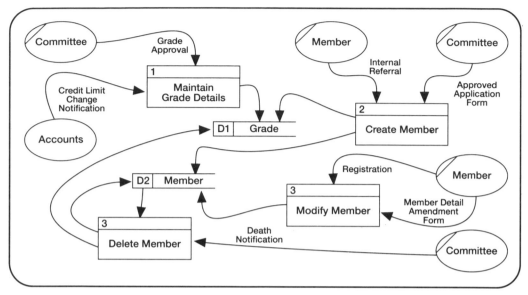

Figure 8.2   Data Flow Diagram for simple 'Club' system

This is a two dimensional grid recording which entities each event affects and the nature of that effect. The columns of the matrix are labelled with the entities from the LDS and the rows of the matrix are the events.

An external event may be recognised on the DFD as the arrival of a data flow from an external source. An internal process or time-based event may be recognised by a process with input data flows from data stores only.

The simple example in Figure 8.1 shows two entities with their associated data items. The Data Flow Diagram for the system is shown in Figure 8.2.

From inspection of the Data Flow Diagram the following events are identified:

— *Approval of a new Grade.* This is clear from the DFD which shows the data flow Grade Approval leading to a process which creates an entity occurrence of Grade.

— *Credit Limit Change.* This is again clear from the DFD with a data flow arriving from Accounts. This shares the same process as Grade Approval but is obviously a different event in the operation of the system. It is quite common for more than one event to share the same process box on the DFD.

— *Application Approval.* An entity occurrence of Member is created either by the receipt of an approved application form or by recommendation from a current Member. This latter flow is called Internal Referral on the Data Flow Diagram.

— *Member changes.* Such as address, name and telephone-no details are notified by the Member and are again clearly identified on the DFD. In this example the membership grade cannot be changed once the member has joined.

— *Member Resignation.* This is also a data flow on the Data Flow Diagram. However, it is necessary for the system to store Member records for seven years before they can be deleted. Hence there is a further event (Resignation > 7 years) and this is time-based. A clue to such events (and internal process events) are that they are usually represented by a read access to a data store on the DFD. In this example, the process Delete Member reads the data store Member. Deletion of member details also takes place on the receipt of a death notification from the Committee. This data flow is also identified as a separate event.

For each event it is necessary to identify which entities are affected. Each effect must be classified as:

- Creation of an entity occurrence (C)
- Modification of an entity occurrence (M)
- Deletion of an entity occurrence (D)

C, M or D is entered into the cell forming the intersection of the event row and the entity column. Some events may have effects on more than one entity, while other events may have a combination of effects on the same entity according to role.

The table below summarises the results of considering each event in detail.

EVENT	EFFECT
New Grade Approval	Creates a new entity occurrence of Grade
Credit Limit Change	Changes the credit-limit data item in a Grade occurrence
ApplicationApproval OR Internal Referral	Increments the no-of-members-in-grade data item in a Grade occurrence. Creates a new entity occurrence of Member
Member Detail Change	Changes member-name, member-address, member-tel-no in a Member occurrence
Resignation Received	Enters Resignation-date in a Member occurrence
Resignation > 7 years OR Death Notified	Deletes a Member occurrence Decrements the no-of-members-in-grade data item in a Grade occurrence.

This understanding is summarised in the Event/Entity Matrix given in Figure 8.3.

Once the matrix has been completed it can be reviewed for completeness. For each column, that is each entity, there should normally be at least one creation (C) effect and at least one deletion (D) effect, unless it has been specified that occurrences of this entity are never deleted. If not all attribute values of the entity are to be stored at creation time then there should be at least one modify (M) effect which establishes these values. If established attribute values are changed during the life of the occurrence this should also be shown by modify (M) effects.

In our example the entity occurrences of the Grade entity are never deleted. This is confirmed in subsequent discussions with the users.

	Entities	
	Member	Grade
New Grade Approval		C
Credit Limit Change		M
Application Approval	C	M
Internal Referral	C	M
Member Detail Change	M	
Registration Received	M	
Registration > 7 years	D	M
Death Notified	D	M

Figure 8.3   Entity/Event matrix for the 'Club' system

8.4   ENTITY LIFE HISTORIES: NOTATION

The Entity Life History is a tree-like structure where:

— nodes are drawn as square-cornered boxes;

— the root node represents an entity type and contains the name of the entity;

— the elementary (leaf) nodes represent events which have an effect on the life of an entity occurrence of the entity type;

— the elementary nodes (effects) contain the name of the event.

The typical life of an entity starts with an event which triggers processing to create a new entity occurrence. Once an entity occurrence has been created then an event can trigger processing whose effect is to modify attribute values of that occurrence. At some later time in the life of the entity occurrence an event will occur which will trigger processing which will have the effect of terminating the life of that occurrence. A terminating event simply means that the occurrence is no longer of interest to the system. This may mean that the occurrence is archived or transferred to another file for use by other systems. This typical life of an entity illustrates the fundamental structure of an ELH.

Sequence

The sequence structure is fundamental to all ELHs. This structure is shown in Figure 8.4. It shows the chronology of events for a Course Attendance. Notification of Enrolment (create occurrence event) always takes place before the Written Examination Result can be notified. The Written Examination Result event occurs before the Notification of Graduation (terminate occurrence event).

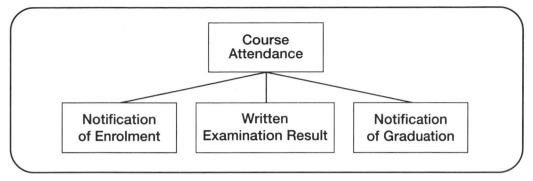

Figure 8.4   Sequence of Events

Selection

In Figure 8.5 the small circles in the two Assessment events denotes that these are alternate events. Thus an assessment event is either a notification of a Written Examination Result or a notification of an Oral Examination Result. The alternative events are grouped under a selection node.

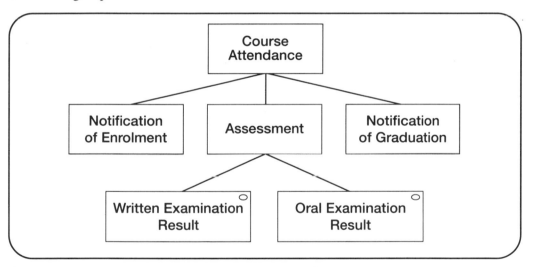

Figure 8.5   Selection of Events

Iteration

The asterisk in the effect box in Figure 8.6 denotes that the event may affect an entity occurrence zero or more times. Continuous Assessment Result may have zero, one or many results before the Notification of Graduation occurs.

Figure 8.7 illustrates the iteration of a substructure of the ELH. In this case the sequence of events notification of Written Examination Result and notification Of Oral Examination Result may not occur, may occur once, or may occur many times. Each instance of the iterated sequence must be complete before the next sequence can begin.

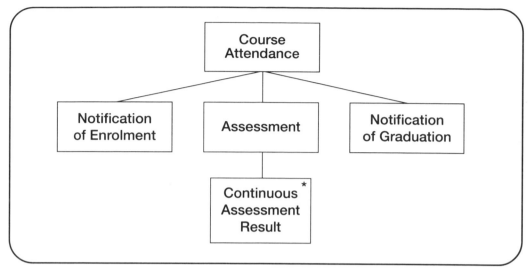

Figure 8.6   Iteration of an event

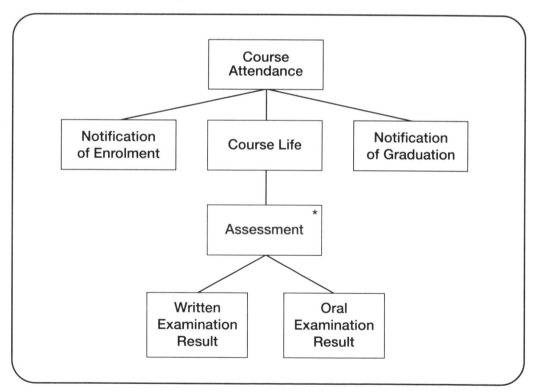

Figure 8.7 Iteration of a sequence

Thus the business rule is recorded that a subsequent written examination cannot take place until an oral examination has occurred, or more usually, every written examination is followed by an oral examination.

8.5 DRAWING THE ENTITY LIFE HISTORIES

The Event/Entity Matrix does not completely represent the life of an entity occurrence. It does not show which events may occur repeatedly, which are alternate events, or the sequencing of those events on a single entity occurrence. Furthermore, the different effects depending on the state of the entity occurrences are also not shown. For this the graphical notation of the Entity Life History is required.

One Entity Life History is drawn for each entity on the Logical Data Structure. Initially only a simple life history is drawn, based on the premise that all occurrences must be created and will probably be modified and will finally be deleted. There will be no consideration of abnormal or exceptional events.

The preferred sequence of completion is to start with the lowest level entities on the Logical Data Structure, that is, entities which are only details, and as each ELH is completed, systematically work on their masters up through the LDS until the top-

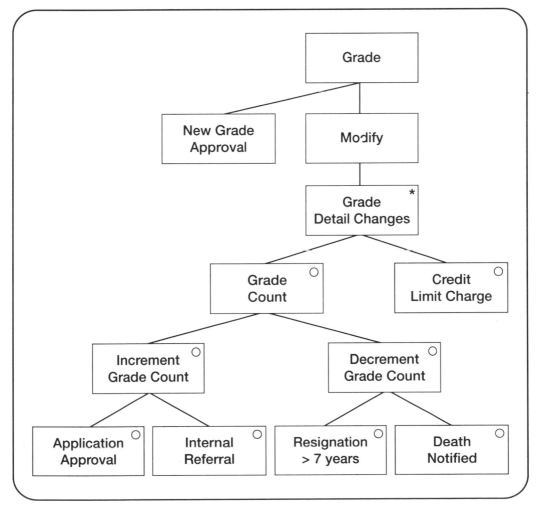

Figure 8.8   Entity Life History for Grade

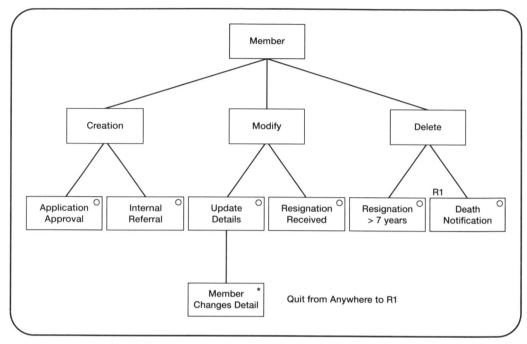

Figure 8.9   Entity Life History for Member

most master entities are reached. This method of working leads to a better understanding of the life of each entity and the effect on its master(s).

The construction begins by selecting from the Event/Entity Matrix all the events which may cause an occurrence of an entity to be created. If there is more than one event then these events are shown as selections under a structure box. It is the creation or birth event which stores a value for the primary key and many of the other attributes.

All the events which cause modification to an existing entity occurrence are now considered. The sequence in which these events will be notified is now decided. Consideration is also given to whether any events occur more than once (iteration), are alternatives (selection) or are optional. A structure is built up using the sequence, selection, and iteration constructs.

Finally the deletion event is added. If there is more than one way that an occurrence can be deleted then a selection structure should be used.

The Entity Life Histories for Grade and Member are shown in Figure 8.8 and 8.9 respectively.

Structures may be combined to represent all the possible lives of entity occurrences as long as the event box remains the elementary node of the structure. This grouping under structure nodes ensures that the different component types are not mixed at the same level within one branch of the structure. Whenever possible structure nodes should be meaningfully named with a group heading which normally defines a type of sub-life or timeframe in which the events occur.

Further notation can be defined for events where there is no agreed sequence (the parallel construct) and for instances where an entity occurrence reverts to an earlier stage in its life or jumps to a later stage (Quit and Resume).

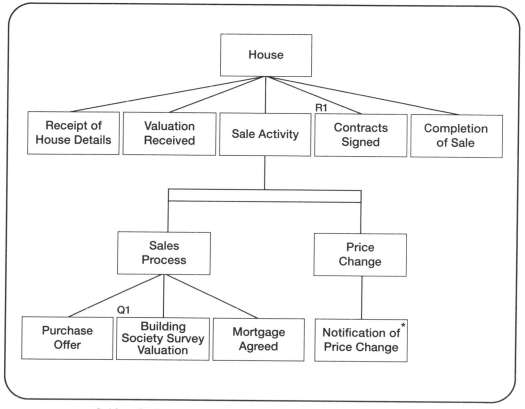

8.10 Quit and Resume and parallel events notation

Figure 8.10 illustrates the use of this notation. It shows that it is possible that following a Purchase Offer for a house the next event may be the Signing of Contracts. This is indicated by the Quit (Q1) and the Resume (R1) notation. However the normal sequence of events is for the Purchase Offer to be followed by the Building Society Survey Valuation. Furthermore the Sale process is independent of a Price Change. Hence a change of price may be notified at any time during the sale activity.

A further variation of the Quit notation is shown in Figure 8.9 where it is possible to quit from anywhere to the Death notification event (R1).

8.6 STATE INDICATORS

State indicators are a method of controlling the sequencing of events. They may be thought of as an additional attribute within each entity. Each time an event affects an entity the state indicator is updated to indicate that the particular effect has occurred. Holding a state indicator in each entity means that it is possible to detect the state a particular entity occurrence has reached within its life. Without a state indicator, complex logic may be needed to investigate many attribute values in different entities.

Knowing the current state of an entity occurrence means that it can be determined whether it is valid to apply the effect of a particular event to an entity occurrence or whether an error condition exists. This implicit validation logic will be carried forward into logical design and built into the processing logic. State indicators also enable selection of entities which have reached a particular stage of their life and so will aid query and report processing. Given that state indicators reflect the structure of the ELH diagrams, their addition to the diagrams is essentially a mechanistic procedure.

Recording State Indicators

For each event the notation used for state indicators is in the form:

valid prior value(s) / set to value

- *valid prior value*(s) gives the values of the state indicator that must exist for the effect of an event to take place.

- *set to value* is the value given to the state indicator once the effect of the event has been completed.

- there may be more than one *valid prior value* of the state indicator for an event.

- there will be only one *set to value* for each event within each ELH.

Set to numbers are written below each effect box on the diagram. The simplest method is to allocate 1 to the birth effect and to number sequentially through the diagram. The termination event or events do not need a *set to state* as the entity occurrence will no longer exist in the system for events to affect it. This is indicated by the null state indicator, a hyphen (-).

Prior values are assigned by recording the state value of all effects which may immediately precede the event. The birth event, or events, will not have any preceding states; this is indicated by the null state indicator.

A sequence of events will have the state indicator set to the value of the preceding effect (*see* Figure 8.11).

Alternatives within a selection structure will each have the same valid prior values. The event following the selection will have as valid prior values all the set to values within the selection structure (*see* Figure 8.12).

Iterated events will include the set to state indicator value set by the last effect of the iteration, as in Figure 8.13. The event following the iterated effect must include the set to value of the effect preceding the iteration as iteration event(s) need not occur.

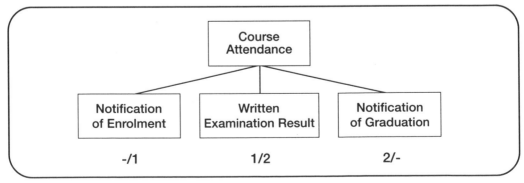

Figure 8.11   State indicators – sequence

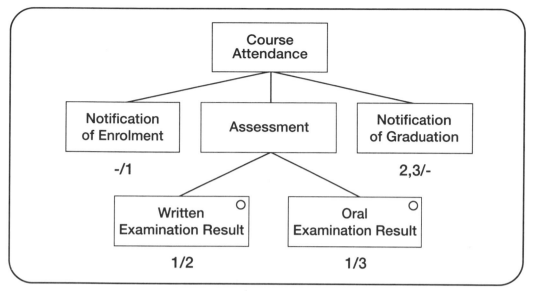

Figure 8.12   State indicators – selection

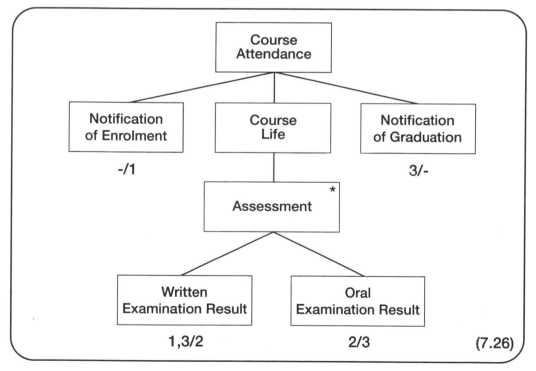

Figure 8.13   State indicators – iteration

The Resume effect must have as a valid prior value the set to value of the Quitting effect, as will the effect diagramatically succeeding the quit (Figure 8.14). Where parallel structures have been used only one of the legs is considered as the main life of

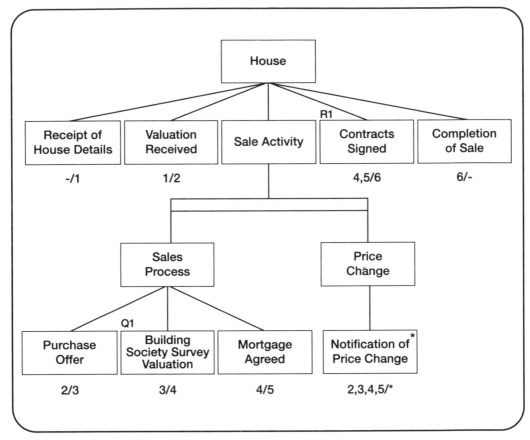

Figure 8.14   State Indicators – Quit and resume and parallel structures

the entity. This is usually the first leg in the structure. It is only the effects on this main leg that may update the state indicator. All other legs of the parallel structure must leave the primary state indicator unchanged. This is indicated by the use of an asterisk (*) as the set to value.

Figures 8.15 and 8.16   show the State Indicators added to the Grade and Member Entity Life Histories.

The addition of the state indicators to Member has also led to the following clarification of the business system requirements;

- Member details will not change once a resignation has taken place.
- A death notification may be received before 7 years has elapsed since resignation.

8.7 SUMMARY

The construction of Entity Life Histories is concerned with:
- identifying events that cause changes in stored data;
- establishing the sequence of these events;
- defining valid states for entity occurrences.

Entity Life History analysis is concerned with eliciting and modelling the sequences and constraints of the business system. The users involved in entity-event modelling must know the business rules, be able to describe unusual business situations, define how to handle error situations and be aware of the way different business areas interact and the movement of data between them.

The detailed analysis of the ELH technique develops further understanding of the data and this will feedback into the Logical Data Structure. This results in adding new entities or splitting existing entities, removing entities as a result of merging, changing relationships or adding relationships, adding or removing attributes.

Events are identified from the Data Flow Diagram. The detailed consideration of these events is likely to lead to amendments of the required logical Data Flow Diagram. It is usual for ELH construction, with its consideration of the effects of

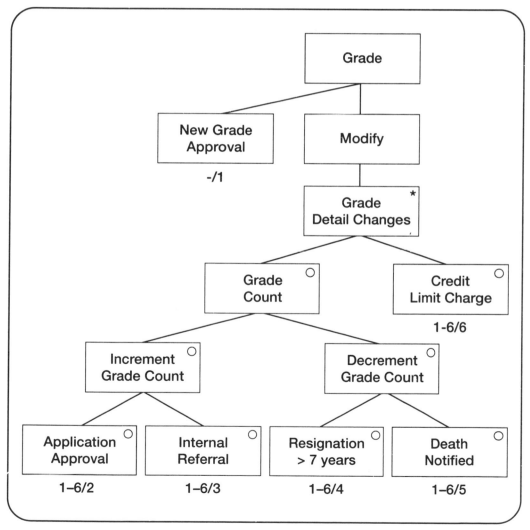

Figure 8.15   Grade ELH with state indicators added

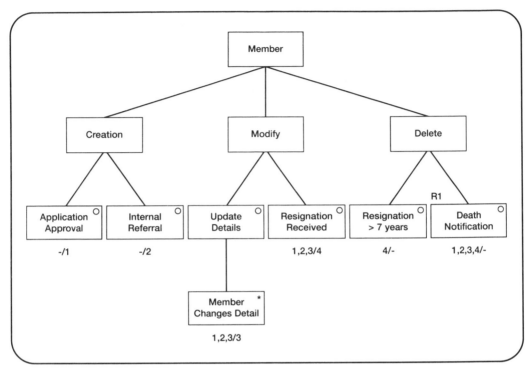

Figure 8.16   Member ELH with state indicators added

events on the stored data items, to produce a much more robust and defensible DFD. The completion of Entity Life Histories directly leads to more confidence in the correctness and interpretation of the Data Flow Diagram, as well as improved understanding of how processes and events affect the stored data structure of the Logical Data Structure.

9 Supporting Models

9.1 INTRODUCTION

This chapter considers three supporting models. Two of these are simple cross-reference grids which should contribute towards the rigour of the main development models. The third, the Data Dictionary, is a fundamental concept underpinning all logical system development.

9.2 ENTITY / DATA STORE GRID

This cross-references the stores of the Data Flow Diagram with the entities of the Logical Data Structure. Except in the most trivial of applications it is impossible to show every entity as a data store on the DFD. Consequently it is necessary to introduce a shorthand whereby one or more entities are grouped together into one data store. The relationship is shown below

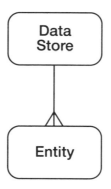

It must be recalled that the grouping of entities is solely for the purpose of clarity in the DFD and in some respects it can be quite arbitrary as long as the cross-reference grid (Figure 9.1) is available. However, there is no suggestion that entities will be combined in this way when physical files are defined later in the development process.

The construction and maintenance of this grid provides a valuable cross-check between the two models and usually leads to a more accurate definition of the Data Flow Diagram.

GRID CHART

| SYSTEM NAME:
Example | | | | DATE:
01/01/91 | | ENTITY/DATA STORE | ✔ | REF NO.
EG2 |
| | | | | | | DATA ITEM/EVENT | | |

DATA STORES

ENTITIES

	Customer	Order	Product	VAT	Invoice	Delivery Note									
Customer type	✗														
Customer	✗														
Order type		✗													
Order		✗													
Order line		✗													
Product type			✗												
Product			✗												
VAT type				✗											
Invoice					✗										
Invoice line					✗										
Location	✗														
Delivery note						✗									
Delivery line						✗									
															AUTHOR: *JM*

TAP 17

Figure 9.1   Entity/ Data Store Grid

GRID CHART

SYSTEM NAME: Example		DATE: 01/01/91	ENTITY/DATA STORE — DATA ITEM/EVENT ✔	REF NO. EG1

	Receive new warehouse details	Receive new product details	Goods delivered to warehouse	Modified warehouse details received	Warehouse delete instruction received	Product amendments received	Product deletion instructions received	Modified location details received	Goods allocated to customer			
WAREHOUSE												
Warehouse-id	C				D							
Warehouse location	C				D							
*Region Code	C			M	D							
PRODUCT												
Product ref code		C					D					
Product type code		C				M	D					
Product description		C				M	D					
Product unit price		C				M	D					
*Product VAT code		C				M	D					
WAREHOUSE/PRODUCT												
Product ref code			C		D		D					
Warehhouse id			C		D		D					
Bin location			C		D		D	M				
Qty in stock			C, M+		D		D	M-				
Qty allocated			C		D		D	M+				
Reorder level			C		D		D	M				
Reorder Qty			C		D		D	M				
										AUTHOR: JM		

TAP 17

Figure 9.2   Entity-Data item/Event grid

9.3   ENTITY-DATA ITEM / EVENT GRID

This grid can be used to show how the values of data items for each entity change as a result of events occurring within the system. It is essentially an extension of the Entity / Event Grid introduced in the previous chapter.

A simple grid is used with entities and their data items listed down the left hand side and events across the top. The effect of each event on the entity data item is shown on the grid intersection as one of the following;

C   Create   The event creates an occurrence of the data item.
D   Delete   The event deletes an occurence of the data item.
M   Modify   The event modifies an occurrence of the data item.

Combinations of C,D and M are allowed. The notation can be annotated as required, for example

M+ to increment data item values.

M− to decrement data item values.

M/F to show a flag field being set to false.

A numeric subscript can be used to show the order in which entities are accessed (not shown in Figure 9.2).

Certain entites are very simple in that all the data items are created by the same event, all (except key values) are modified by the same events and all are deleted by the same event. It is unnecessarily cumbersome to list all the data items in such circumstances and the grid may be reduced to a simpler Entity / Event matrix.

9.4   DATA DICTIONARY

The Data Dictionary holds the corporate data resource and provides a detailed underpinning of the other models introduced in this text. A Data Dictionary is simply a record of data about data. It may be manually compiled (on A4 sized sheets in a loose-leaf folder) or it may be a fully automated package. There are considerable benefits to

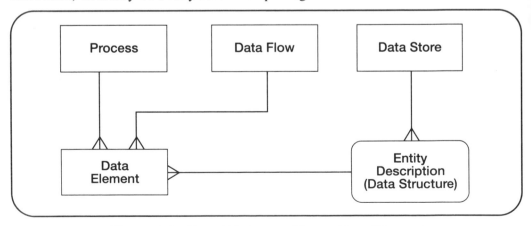

Figure 9.3   Data Dictionary Entry: Data Element

be gained from using an automated dictionary and these will be discussed later in this chapter. However, for the present purpose of illustration, a simple manual system will be assumed.

It is usual to hold data about three of the four main constituents of the Data Flow Diagram. These are the store, the process and the flow. It is also necessary to record information about the entity descriptions that support the Logical Data Structure and the attributes (data elements or data items) contained in each description.

These five Data Dictionary types are related in the following way: (see Diagram)

Thus it is necessary to hold Data Dictionary entries about data elements, data structures, data flows, data stores and processes. The structure of the Data Dictionary for each of these will vary and Figures 9.3 – 9.7 are meant to be illustrative.

*Data Dictionary Entry: Data Element (**Figure 9.3**)*

Name:
A meaningful unique name

Description:
A short description of the meaning of the data element. An example might be included.

Aliases:
Several departments may refer to the element by a different name or term. This has to be explicitly recognised and may require a large amount of detective work. Different employees in an examination system used the terms "unit" and "module" to refer to the same data element. This was only recognised through the analysis work required in the compilation of a Data Dictionary.

Type:
Usually Character, Numeric or Alphanumeric.

Format:
To prepare for Format checks in the subsequent system design. A convention of representing numbers by nines (9) and characters by Xs can be adopted.

Values:
Discrete data elements have a meaning associated with each value. Listed below are some examples from a payroll system

033	Arrears of Pay	016	Honorarium
701	Unpaid Leave	710	Sickness Benefit
052	Maternity Pay	119	Split Duty Payment

Data about ranges and discrete values provides the basis of many of the data validation checks required in the subsequent system.

Security:
Who (or which level of employee) is allowed to modify, add or delete a given data item. This will be important in the design of security features such as passwords and audit checks.

Editing:
This may concern the way in which the data is produced from the system. For example, should a credit of £30 be shown as −30, +30, 30CR or (30)?

Comments:
A final section in which to record special information about this data element.

DATA ELEMENT NAME: _APPL-REF_

SHORT DESCRIPTION _A unique code given by the college to each_

applicant applying for a full-time course (Application-Reference)

ALIASES _Applicant-number_   TYPE _Alphanumeric_

FORMAT _XX99999_

─────────────────────── VALUES ───────────────────────

DISCRETE	CONTINUOUS
NOT applicable	*LP00001 - LP99999 (this year)*
	XX00001 - XX99999 (general)
	Note: unlikely for values to exceed
	LP20000

SECURITY   *Created only by admissions tutor*
Deleted only by admissions tutor
May be viewed by all staff members

EDITING

Not Applicable

COMMENTS

Serial number. Two character prefix changes from year to year

Figure 9.3   Data Dictionary: Data Element

DATA STRUCTURE NAME: _APPLICATION_

SHORT DESCRIPTION _Describes data elements associated with a_

person applying for a course place

CONTENT *APPLICANT:*

APPL-REF	SEX
SURNAME	MARITAL-STATUS
FORENAME	BIRTH-DATE
DESIGNATION	BIRTH-COUNTRY
[PREV-SURNAME]	BIRTH-RESIDENCE
HOME-ADDRESS	NATIONALITY
[CORR-ADDRESS]	{ PARENT-NAME
[HOME-TEL NO]	GUARDIAN NAME }
[CORR-TEL NO]	COURSE APPLIED* (1-5)

VOLUME INFORMATION

2000/Year

COMMENTS

Figure 9.4   Data Dictionary: Data Structure

Collecting and recording information about data items will begin early in the systems development process. The data items described in the Clerical Document Specification (*see* Chapter 4) are clearly the basis of Data Dictionary entries that will be developed as analysis proceeds.

Data Dictionary Entry: Entity Description (Figure 9.4)

An entity description is a set of related data elements. As analysis progresses it is likely that this set will become well normalised (*see* Chapter 6). The detail of each element is documented elsewhere in the Data Dictionary (*see* above). The list may also contain data elements which are either optional, repeated or mutually exclusive. The following convention can be adopted:

Optional Structure: Placed in square brackets.

[PREV-SURNAME]

Alternate Structure: Placed in braces.

{PARENT-NAME}
{GUARDIAN-NAME}

Iterations of structure: Marked with an asterisk.

COURSE-APPLIED* (1-5)

With the number of iterations placed, if known, in parentheses. In this case an Applicant can apply for 1,2,3,4 or 5 courses – but no more.

Volume information collected at the end of the form will be needed when the size of the system is estimated.

Data Dictionary entry: Data Store (Figure 9.5)

These are defined in terms of entity descriptions. The concept of the entity/ data store grid has been introduced earlier in this chapter. Hence each entity is allocated to a store of the Data Flow Diagram. It may also be possible to include the Search criteria in the definition. An example is given illustrating how the request Applicant-Status-Query (the current status of the Applicant's application) is always searched for on Appl-Ref – the reference given to him or her by the College.

Data Dictionary Entry: Data Flow (Figure 9.6)

Data Flows are defined in terms of data elements and (where appropriate) entity descriptions. The source and destination of each flow will be given using the reference numbers allocated to the processes in the Data Flow Diagram. Volume information will assist in system sizing as well as in the calculation of input and output times.

Data Flow entries define input and output requirements. They should determine the content of forms and reports as well as guiding the technology used to capture or produce them.

Data Dictionary Entry: Processes (Figure 9.7)

This issue has already been addressed (*see* Chapter 7) where Decision Tables, Decision Trees and Structured English were suggested as appropriate process definition models.

DATA STORE NAME: *APPLICATION*

DESCRIPTION *Details about the current status of all those applicants*

currently applying for admission

DATA FLOWS IN

    EDUCATIONAL - REFERENCE   *WITHDRAWALS*
    VALID-APPL-FORM-A   *FINAL ACCEPTANCES*
    OFFER - DETAILS

CONTENTS

    APPLICANT

    EDUCATIONAL-REF-ASSESSMENT

    CURRENT-STATUS

DATA FLOWS OUT

    STATUS-SUMMARY-REPORT

    APPLICANT-STATUS-QUERY (APPL-REF)

    MISSING-EDUCATIONAL-REFERENCE-REPORT

PHYSICAL ORGANISATION

COMMENTS

Figure 9.5   Data Dictionary: Data Store

DATA FLOW NAME: _APPLICATION-FORM-A_

SOURCE REF: _Source_ DESCRIPTION _APPLICANT_

DESTIN REF: _1_ DESCRIPTION _CHECK STATUS_

DATA FLOW DESCRIPTION _Describes an application form submitted by an applicant_

CONTENT:

COURSE

APPLICANT

APPLICANT - EDUCATION

APPLICANT - SPECIAL - NEEDS

APPLICANT - PAYMENT - METHOD

REFEREE

VOLUME INFORMATION

2000/Year

COMMENTS   APPLICATION FORM - B 15 The EDUCATIONAL REFERENCE. This will accompany PART A if the flow occurs before the 16th July

Figure 9.6   Data Dictionary: Data Flow

PROCESS NAME: _1.1 Determine Nationality Status_   DESCRIPTION: _To determine whether student is classed as specified (overseas)_

INPUTS	LOGIC	OUTPUT
APPLICATION-FORM -A	_If Birth-Country = 'England' or 'Scotland' or 'Northern Ireland' or 'Wales' or Birth Residence = 'United Kingdom' and nationality = 'British'_ _then mark the Application form 'Valid' and Pass to relevant Admissions tutor_ _else mark the Application form 'Specified' and undertake Examination checks_ _ENDIF_	_VALID-APPL-FORM-A_ _SPECIFIED-STATUS -DETAILS_

COMMENTS _Specified students must be subject to examination equivalence checks (1.2) and stringent financial verification (1.3)_ | REFERENCE (FULL LOGIC DESCRIPTION)

Figure 9.7   Data Dictionary: Process

Further possibilities are examined in the companion text on System Design. These process entries form the basis of clerical procedure documents, computer programs or some combination of the two.

In summary, the Data Dictionary represents an important corporate resource it is used as a foundation for the logical system definition. It must be stressed that it is not a static mechanism and it will be built up over a period of time as the analyst gains greater understanding through fact finding.

The Data Dictionary has to support a complex web of inter-relationships. A single data element may appear in many data flows, data stores and processes. The effect of amending a data element has to be 'traced' through the system to identify possible problems and side-effects. The automatic maintenance of these links is clearly necessary in all but the most trivial of applications and hence the importance of Data Dictionary software.

9.5   DATA DICTIONARY SOFTWARE (DDS)

Data Dictionary software (usually referred to by the term Data Dictionary in the rest of this chapter) can effectively and efficiently maintain a large central repository of data about the data of an organisation – so-called metadata.

Metadata is a level of abstraction higher than the actual data used in operations; it is not the actual data used but data about the actual data used. Figure 9.8 shows the distinction between the two levels of data. Metadata is used to define, identify and describe the characteristics of the user data. Metadata usually falls into two categories:

- What the data is or what it means.

- Where the data can be found and how it can be accessed.

There has recently been an increased acceptance of the benefits of establishing and maintaining a Data Dictionary and a number of products are available in the

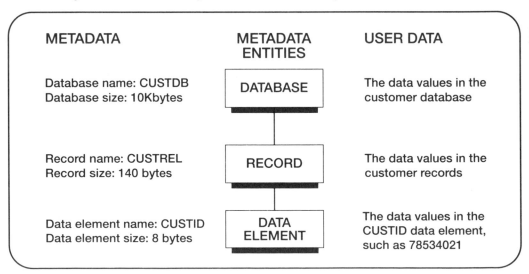

Figure 9.8   Data and meta-data

commercial marketplace. Furthermore, organisations are increasingly recognising that data is an important resource which, like other resources such as staff and materials, has both value and cost to the enterprise.

Consequently data needs to be managed so that it effectively serves the information systems of the organisation. For this management to be successsful there needs to be:

• Knowledge of what data exists and how it is used.

• Control of modifications to existing data or processes using data.

• Control over plans for new uses of data and over the acquisition of new types of data.

(BCS, 1977)

This need for control is increased where data is used for more than one application and the trend towards database solutions naturally means that this is often the case. In such circumstances the central definition of data and its use is usually delegated to a Data Administration function. The Data Administrator will be concerned with the correct use and maintenance of data, the integration of new applications and amendments, and the implementation details of data storage, access and manipulation. The Data Dictionary is an essential support tool for the successful performance of these tasks.

The first automated Data Dictionary Systems were primarily concerned with documenting the physical aspects of data processing – systems, programs, files and databases. However, their scope has now increased to storing and manipulating logical models – such as Data Flow Diagrams and Logical Data Structures and so the Data Dictionary has progressed from a passive documentor of systems to an active productivity tool.

The activity of a Data Dictionary System demands scrutiny. All data processing applications require metadata to operate (such as database schemas, COBOL File Definitions and Job Control Languages) and an active Data Dictionary controls this processing environment. Indeed it is the scope of the metadata management that determines the activity of the Data Dictionary because the DDS is only active "with respect to a program or process if and only if that program or process is fully dependent upon the Data Dictionary for its metadata" (Plagman,1978). In a passive system metadata may be defined from other sources and the Data Dictionary acts as a documentation facility rather than an active tool in system development. In general, commercial DDS may be placed upon a spectrum of activity with the trend towards active or potentially active Data Dictionary Systems.

This concept of activity is an example of how the term Data Dictionary is used by vendors to describe software with a wide range of capabilities and facilities. Comparisons across products reveal marked differences in functionality. Consequently, the next section describes desirable features of a Data Dictionary system with the aim of providing an appreciation of the scope of DDS as well as a framework in which to assess competitive products. It is largely based upon the BCS Data Dictionary Systems Working Party report of 1977 which still remains an important benchmark in this field.

9.5.1 Desirable Features of a Data Dictionary

The BCS Working Party recommended that the Data Dictionary should operate at two distinct levels. The first is the logical level that gives the ability to record and analyse requirements irrespective of how they are going to be met. This logical view represents an implementation independent view of the enterprise and initial and successive implementations must take place within the scope of this framework. This book has concentrated on the logical definition of the system.

The second level of the Data Dictionary is the implementation level. This gives the facility to record physical design decisions in terms of the implemented database or file structures and the programs that access them. If the logical level is how the data is seen from the enterprise, then the implementation level is how the data is viewed by the file handling system or the DBMS itself.

Logical Level

The logical view describes the nature of the enterprise and its data. It is a model of the organisation showing things of interest to it, functions it can perform and events which influence the way it performs. It is independent of any current or proposed implementation and so represents the logical requirements that successive implementations are designed to fulfil. The Data Dictionary should be able to support this model.

At this level the Data Dictionary should be able to record details of

— entities and relationships of concern to the enterprise;
— processes of interest to the enterprise or carried out by it;
— responsibility for processes, perhaps in terms of the structure of the organisation;
— flows which result from processes or from external entities or events;
— the connections that exist between entities, processes and events.

The Data Dictionary should be able to record details of different versions recognised as valid at different times or contexts. It is also essential that the Dictionary can define the relationships between the logical entities and the corresponding files and records of the implementation.

Implementation Level

The implementation view is the basic source of information about the physical data processing system. It provides data to help establish the design of the system, to prove its correctness and to identify the impact and cost of changes. It is likely to represent a flexed implementation of the logical model. It must be logically consistent with this latter model and also not exceed it in scope. At this level the Data Dictionary represents a coherent, centralised library of data about all aspects of the data processing system, enabling all users to have a clear and consistent view.

Two examples of such data are:

• *Data Description elements.*
  These will describe the different data types and structures used in the system, such as records and files. Elements should be described in terms of their:

— Names: including aliases and past names.

— Classification: Description, ownership, status etc.

— Representation: Type, length, order, etc.

— Use: Frequency and volumes.

— Administration: Memory and storage requirements.

• *Process Description elements.*
These will also demand the same type of metadata as the Data Description elements. Further information might include:

— Program size – in some appropriate metric.

— Processing Type – Batch or On-line.

— Parameters – Number and types required.

Several versions of programs and data structures may exist at any one time. These may represent live, test or design states and this must be recognised and recorded as such.

The Data Dictionary should also enable descriptions of the implementation level structures to be established and maintained. This may be achieved through a direct input language, from program data definitions in high level languages, from a DBMS source definition or from program procedure definitions.

The implementation view also demands details about the physical storage of data and its use. Facilities required include recording of physical attributes such as

• Storage media   –   storage media type e.g. disk.

• Storage size   –   describes space requirements.

• CPU   –   describes the CPU name and size required.

The Data Dictionary System should validate input for syntax, consistency and completeness. These checks should include:

— the characteristics of each physical file;

— the contents of each file;

— each physical structure. Checks that all the constituents of the implementation data structure are allocated to at least one physical file.

The implementation view contains all the information necessary to derive an 'optimum' operational schedule. This is supported by the collection of performance and utilisation statistics such as.

• Frequency   –   indicates the average frequency that the file is accessed (such as daily, weekly).

• Response   –   refers to the response time of a process.

• Log information   –   shows statistics on when a record or file is accessed, by whom, and the activity that is performed.

• Usage statistics   –   records summary of usage.

Holding data volumes in each operational definition and the physical description of the files themselves provides information for the realistic simulation of database performance. This gives the facility to tune the performance of the system to achieve a database or file structure that gives optimum performance.

9.6 CASE TOOLS

The last decade has seen the emergence of CASE (Computer Aided Software Engineering) tools to support the structured development life cycle. The data dictionary (encyclopaedia or repository) is a fundamental construct of the CASE tool. However, CASE also offers:

- A diagramming 'front-end' to describe the system using appropriate logical models. Examples of these models; logical data structures, data flow diagrams and entity life histories have been described in this text.

The CASE tool may offer

   — a graphical editing facility to produce quality outputs which can be quickly produced and amended;

   — a completeness check for each model to ensure that it is logically correct;

   — a consistency check between models to ensure a common consistent view of the system.

- A prototyping facility for inputs and outputs based on the logical data contents defined in the data dictionary entry supporting the logical data flows.

- A program and data generator converting logical process definitions into programs and data stores/entities into physical files, tables or databases.

Most traditional Data Dictionary vendors have extended their products into offering CASE. Some of the important functions of a DDS/CASE tool are summarised below.

Consistency Checking

This is an essential feature of a CASE tool. In the context of a Data Flow Diagram this can answer such questions as:

— Are there any data flows specified without a source or destination?

— Are there any data elements specified in any data stores that have no way of getting there, as they are not present in any of the incoming data flows?

— Does a process definition demand a data element that does not enter that process?

— Are there any data elements in any data flows entering processes that either are not used in the process and/or do not appear in the output?

The verification of system consistency is a vital task that eliminates a considerable amount of desk checking. It permits the insertion of vital, but omitted, data elements and the deletion of irrelevant ones. Consequently, the CASE tool not only ensures the validity of the design but also identifies and justifies the role of each data element.

Thus it is possible to demonstrate why certain data is collected and where and how it is used.

Consistency may be checked within each model and between models. This leads to consistent and complete specifications well in advance of the production of program code.

Testing

The development and entry of volume test data is very time consuming. However, if descriptions and value ranges are already stored in the dictionary then test data can be automatically generated. A randomising facility can produce erroneous data which should be trapped by the testing.

Coding

In most CASE tools the description of the data structures in the data dictionary should be detailed enough for the generation of data descriptions in the host language or Data Manipulation Language (DML) through a pre-compilation pass of the dictionary. In fact some CASE tools are offered by the vendors of Data Base Management Systems (DBMS) to permit the definition of data files through a diagramming front-end.

If the CASE tools holds information on the logical sequence in which data is used by processes, then automatic program code definition is feasible. CASE tools which offer this facility should give significant productivity gains, not only because of the speed with which code is produced but because of the quality of that code. Program testing should not be necessary. A review of four early users of Information Engineering Facility revealed systems with six million lines of COBOL code going into production without a single failure.

Change

The tracing of the effects of required changes through the complete system is a time consuming and difficult task. Impact Analysis is the term given to the analysis of the effect of proposed program and system changes. The recording of the relationships that exist between the various entities should allow the effects of addition, amendment or deletion of a particular entity to be predicted throughout the whole system. Thus the system and resource implications of a change can be completely understood before they are implemented.

In an active dictionary it may be possible to generate some, or all, of the changes automatically. This is particularly true of data validation changes where the rules for validation (for example, format, length and value ranges) are maintained in the Dictionary, not the program code. Thus changes only have to be defined in one place – the Dictionary entry – not in all the programs that access that data.

In circumstances where the system has been automatically generated from a CASE tool, changes to the system are defined in the documentation of the logical model not the code itself. Once the changes have been checked for completeness and consistency the whole software system is re-generated.

Reporting and documentation

The initial documentation role of many Data Dictionaries has led to most DDS and CASE tools having flexible and comprehensive reporting facilities. A variety of analyses will be required and the facility to search for textual descriptions is essential if the full potential of the system is to be realised.

In a fully integrated CASE tool the documentation is central, because it is "where the code comes from, where it is tested and where it is maintained". In the CASE approach the documentation is the system, not a partially complete afterthought.

9.7   DATA DICTIONARY: SUMMARY

Thus the Data Dictionary is an important tool in managing the corporate data resource. Furthermore, it underlies all the models that have been introduced in the rest of this text, providing the metadata detail required for a successful understanding of current practices and a sound design of their successors. Data Dictionary Systems have progressed from documentation tools performing basic cross-references to software that provides support for all stages of systems development.

Current Data Dictionary Systems are more active than their predecessors. In a number of cases it is the Dictionary that drives data analysis and database design tools, application generators and report writers, with the dictionary facility being used to co-ordinate these various tools, models and phases. CASE tools rely upon a central dictionary or encyclopaedia. CASE offers diagramming of common logical models and significant quality and productivity gains through automatic code generation. Some CASE tools support a particular methodology whilst others may be tailored to reflect a company's preferred approach. These latter meta-CASE tools reduce the actual and hidden costs often associated with moving from current practices to a 'standard' methodology.

10 Other Perspectives

10.1 INTRODUCTION

This chapter examines four further perspectives:

- An extended Data Flow Diagram to reflect the requirements of modelling real-time systems.

- An alternate data modelling convention using the Chen notation.

- The principles of prototyping.

- The emerging concepts of Object-Oriented Analysis (OAA).

10.2 DATA FLOW DIAGRAMS (DE MARCO, WARD & MELLOR)

Tom de Marco's seminal text has been referenced many times in this book. For many of us reading it at that time it threw light on activites that had previously seemed chaotic, ill-defined and amateur. His original notation for Data Flow Diagrams is shown in Figure 10.1 alongside contemporary SSADM equivalents. It is his set that was extended (Ward and Mellor, 1985) to allow the modelling of real-time systems. Real-time systems are generally process rather than data driven and this emphasis is reflected in Ward and Mellor's emphasis on extending flow and process notation.

This short section introduces some of the modelling notation used in this real-time extension by re-examining the constituent parts of the DFD.

Data Flows

The data flow symbol is extended to distinguish between discrete and continuous flows and between data and event flows. A time-continuous flow is defined as a flow that exists at every instance in a time interval and it is shown on the DFD as a data flow with a double arrowhead. For example, a process concerned with maintaining temperature may continually sample the temperature of the surrounding air (*see* Figure 10.2). In contrast a discrete flow only has values at certain points in time and at all other times it is considered to be null or undefined. Thus the time discrete flow is very like the data flow of a business system.

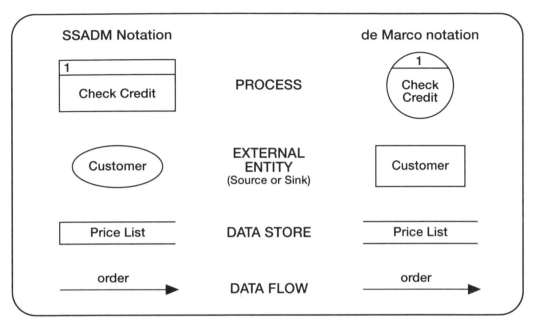

Figure 10.1   Data Flow Diagram symbols

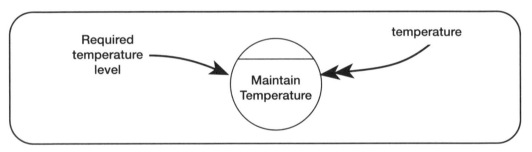

Figure 10.2   Continuous and discrete flows on a DFD

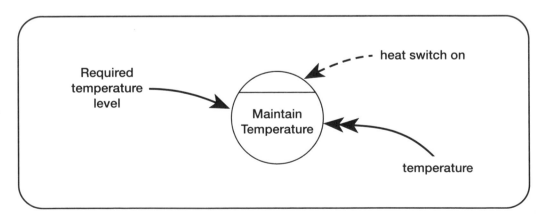

Figure 10.3   Event flow on a DFD

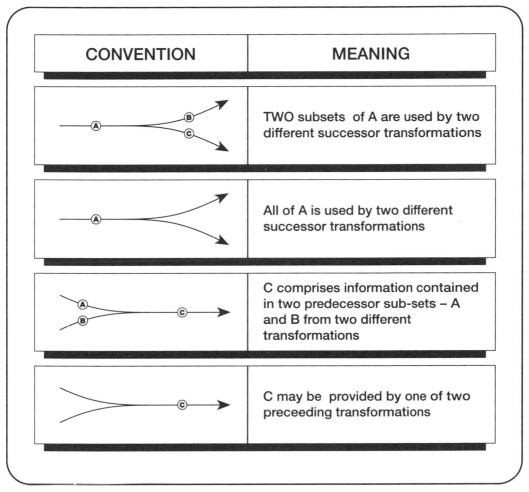

CONVENTION	MEANING
	TWO subsets of A are used by two different successor transformations
	All of A is used by two different successor transformations
	C comprises information contained in two predecessor sub-sets – A and B from two different transformations
	C may be provided by one of two preceeding transformations

Figure 10.4   Convergence/Divergence conventions

Data flows represent both the event (the receipt of an order) and the content of the flow (order details). In contrast an event flow is a flow that has no data content. It is usually a signal to indicate that something has taken place or a command has been issued. The depression of off-switch gives event information to the system but in itself contains no data values. These event flows can be shown on the DFD as dotted lines (*see* Figure 10.3).

Ward and Mellor also permit the convergence and divergence of data flows. This allows the modeller to distinguish between a data flow which is completely or partly used by different successor transformations and vice versa (*see* Figure 10.4).

Processes

A process that only has event flow inputs and produces only event flow outputs is known as a control transformation. It is represented on the DFD as a dotted circle (*see* Figure 10.5). The preferred method of the logical description of a control

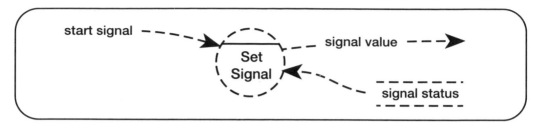

Figure 10.5   Control transformations and event data stores

transformation is the State Transition Diagram. This is examined in detail in the companion text Introducing Systems Design.

Stores

An event store can be defined with parallel dotted lines (*see* Figure 10.5). It holds the occurrences of event flows such as Stop/Start, Red/Green/Amber, etc. Unlike DFDs defined for business systems it is diagrammatically convenient for an event flow to directly pass into an event store without passing through a transformation.

10.3   ENTITY-RELATIONSHIP MODELLING USING THE CHEN NOTATION

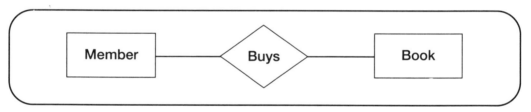

In this notation rectangles are used to display entity types and relationship types are described within diamond shaped boxes. The connecting lines show which entity types are associated by each relationship type. The naming of the entities is usually relatively straightforward, but the relationship types often pose more of a problem. Troublesome relationships may be named by reference to the entities that are joined together.

In the construction of this example we will assume that no multiple copies of books are stocked by a book club, i.e. that the ISBN is the identifier of Book. In practice this will be unlikely, the Book Club is almost certain to hold many copies of popular texts in stock. The implications of this assumption will be explored after the terminology and notation have been introduced. Furthermore, the Member is not identified by name, there will be many John Smiths, but by the membership number allocated by the Book Club.

10.3.1   Properties of relationships

Degree of relationship

One of the most important properties of a relationship is the degree of the relationship that exists between the entity types. In the example used so far a relationship was identified between Member and Book. What is the degree of this relationship? Three possibilities exist and should be considered by the analyst.

— *One:One relationship (1:1)*

This would find expression in the following rule

A Member may buy, at most, one Book.
A Book may be bought by, at most, one Member.

The example below illustrates this rule.

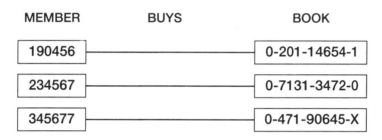

The 1:1 relationship permits a Member to buy only one book at a time and hence is unlikely to be relevant to this example because a Book Club would seldom wish to be so restrictive in their sales practices. However, this is not to say that these circumstances would never occur. Certain organisations dealing in collectors items may wish to record the person who bought particular books and in such an application the 1:1 relationship would be applicable. Member-Numbers would be allocated to members for each sale and so either the Member-Number or the catalogue identifier would be a sufficent identifier for the relationship. The Member-Number could be used for finding the book bought by a certain Member, whilst catalogue identifier would permit the tracing of a specific book. However, another part of the model must recognise that a given Member could have many Member-Numbers to reflect different transactions.

— *1:Many relationships (1:N)*

This covers the following rule:

A Member may buy many Books.
A Book may be bought by, at most, one Member.

An example is again shown overleaf.

This is probably nearer the circumstances of the collectable books example. It is expected that the club would allow a Member to buy many books, but a specific copy of a book may only be bought by one Member. In this instance the unique identifier for the relationship would be the ISB Number. This works in the collectable books example because there is only likely to be one copy of each book. However, this still does not match the rules of the Mail Order Computer Book Club.

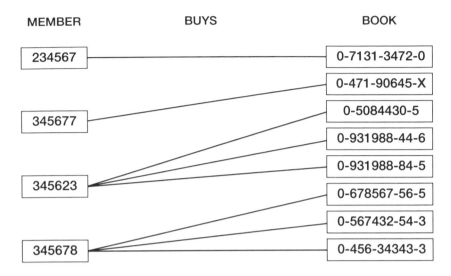

— Many: Many relationships (M:N)

The final type of relationship covers the rule:

A Member may buy many books.

A Book may be bought by more than one Member.

This many to many relationship is written as M:N and is illustrated below.

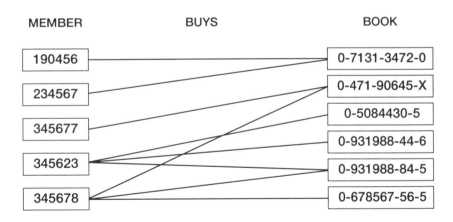

Consider the more likely situation where the club stocks more than one copy of a particular book so that the entity occurrence identified by ISB Number 0-471-90645-X, of which the club has several copies, could be bought by different Members. This could be expressed as a many-many relationship, but in truth it is an indication of a poor entity-relationship model. The entity identifier now fails to uniquely identify a single entity occurrence. The model can be improved by introducing the entity COPY with an identifier Serial-number, where a Serial-number uniquely identifies a copy of a book. So now members can buy copies of a particular Book.

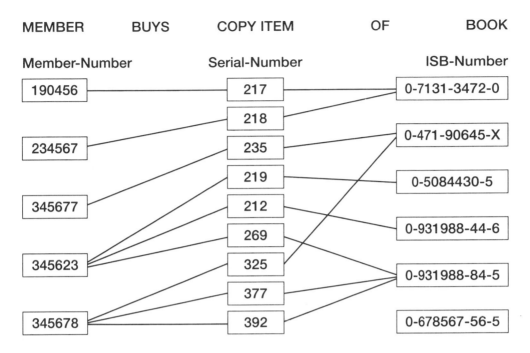

This illustrates how a data model will change as its implications become clearer during the modelling exercise. Although the above example of an M:N relationship is contrived there will be many instances where such relationships will occur naturally.

The notation extended

The degree of the relationship may be shown on the diagrams using the following convention:

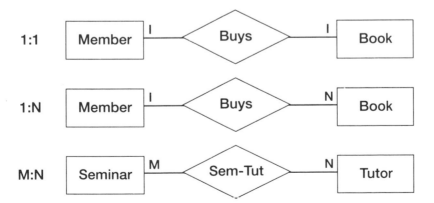

Notice the naming convention of the relationship in the last example. It is sometimes very difficult to give short meaningful names to every relationship. This particular example could be PRESENTS or HAS but it is difficult to identify a statement that satisfactorily describes the two-way nature of the relationship. A tutor presents a seminar but a seminar can hardly be said to present a tutor. In such instances it is much easier to give a name derived from the two connected entities and this convention can be used to describe relationships where no obvious term springs to mind.

Membership Class

A further property that repays investigation is that of membership class. This is concerned with how the nature of the relationship is affected by obligatory and non-obligatory rules found in the problem under investigation. This should become clear by examining another relationship that is likely to occur in the Mail Order Book Club system, that of BOOK and PUBLISHER.

This has the following rule:

> A publisher may publish many books.
> A book is published by, at most, one publisher.

The following diagram summarises this rule:

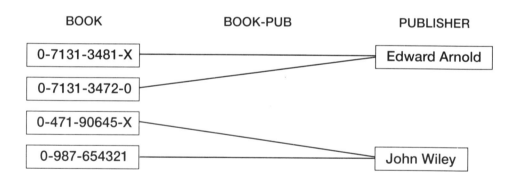

It is possible to extend the Analyst's knowledge of this data by considering the obligatory nature of the relationship. Howe (Howe, 1983) includes this additional information on the data model by using a blob or dot inside the stripe on the entity symbol to signify that the entity's membership class is obligatory. A blob or dot outside means that it is not obligatory.

This diagram now expresses the rule:

A book must have at least one publisher.
A publisher must publish at least one book.

Thus a book without a publisher is not permitted (how does the system cope with internally produced reports?) and a publisher for whom the club has no books is also disallowed.

This shows the rule:

A book need not have a publisher.
A publisher need not have published a book.

This permits the Analyst to include internally produced reports in the model and also to record publisher details even where no books are currently stocked from that publisher. An example for this is shown below.

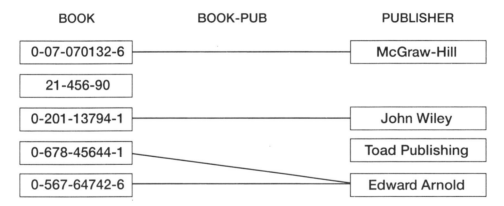

Thus book number 21-456-90 does not take part in the relationship BOOK-PUB because it is internally produced. Similarly there are no books stocked from Toad Publishing.

This covers the rule:

A book need not have a publisher.
A publisher must publish at least one book.

Illustrated by:

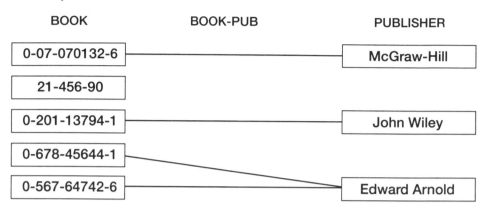

In this instance the publisher Toad Publishing is not permitted to take part in the model of the system.

This diagram covers the following circumstances:

A book must have at least one publisher.
A publisher need not have published a book.

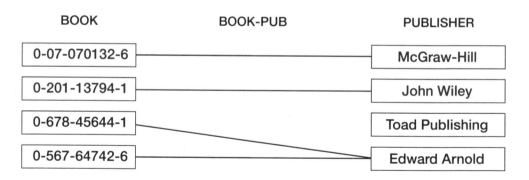

In this model the internally produced report, given the number 21-456-90, is not permitted because it does not have a publisher. However, Toad Publishing does now appear in the model. Knowledge of such classes will come in useful in the design of the file structures.

Many to Many relationships

To examine the breakdown of an M:N relationship consider the relationship occurrences of the following example:

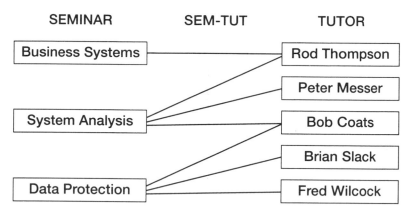

The decomposition is performed by treating the assignment of tutors to seminars as a separate entity. This may appear to be somewhat arbitrary – after all an entity was defined as a thing identifiable to the enterprise – but it is quite legitimate in the model. Things of importance to the modelling of the application may not be evident to those working in the actual operations of the enterprise. The new entity-relationship model for the project example is given below.

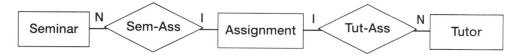

This model is illustrated by:

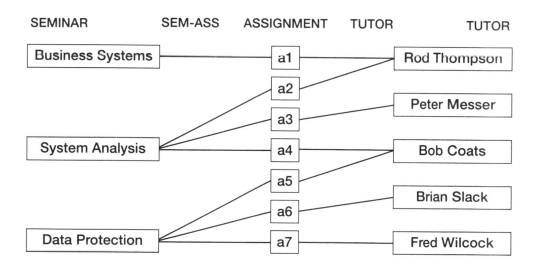

Note that the new entity will have obligatory membership with both of the newly formed relationships. However the original entity's membership class with the new relationship will be exactly the same as it was with the one that has been replaced.

10.4 PROTOTYPING

A major criticism of the Life Cycle model is the time-scale associated with its linear progression of activities. The gap between specification and delivery is often so long that requirements change dramatically in this time. This leads to the delivery of systems that "were required two years ago".

An alternative approach, known as Prototyping, has claimed advantages in most aspects of the Life Cycle. Prototypes are a first attempt at a design which is then extended and enhanced through a series of iterations. Prototyping stresses the early delivery of an incomplete, but working, system and the use of prototypes may be valuable at various stages of the Life Cycle.

Prototyping has clear precedents in other engineering disciplines and activities. It is a live working system and not just a paper based design. Users can test its operations and explore its facilities and so do not have to rely upon written descriptions.

Prototyping is an iterative process. The first system is built around the user's basic requirements and refined in the light of comments and difficulties. Thus the system passes through a number of iterations until it becomes an acceptable reflection of the user's requirements. At this stage the designer has two fundamental choices.

- To refine the prototype into the final system. This will require some development of error trapping and recovery routines. Re-programming of 'working' code may also be needed to reflect internal standards of naming and modularity. The prototype may have been constructed using value rather than parameter passing and temporary data files. These areas will obviously have to be re-addressed.

- To re-code the entire system. The prototype may have been developed with the prime aim of defining user requirements. The software used to do this may turn out to be completely inadequate beyond a certain volume of transactions. Thus re-coding is required to permit the phasing out of the Prototype. Furthermore the target language of the final system may be different from that used to construct the prototype.

It must be clear that, for such a development method to work, the system has to be written in very powerful and flexible software that permits programs to be created very quickly as well as allowing easy coding of extensions and amendments. Conventional programming languages (such as COBOL and BASIC) are not suited for this type of development. Prototyping requires powerful software tools that allow the relatively inexpensive buiding of systems which might eventually be discarded.

Prototype tools

The tools chosen for prototype development must, above all, permit the quick development of working systems. Four candidates may be identified.

- Application Packages

  It may be possible to develop a demonstration system using an appropriate package and to let the user identify problems, possibilities and opportunities using the package as a yardstick. It is often easier to say what is inadequate about, say a production control package, than it is to define requirements in the abstract.

- Program Generators

  Program Generators have been available for a number of years and a variety of different types may be identified. One sort uses a question and answer English dialogue to produce the program logic which it then encodes in a High Level language such as BASIC or COBOL. A different type adopts a screen based approach where the designer effectively **paints** the screen he wishes to be displayed by typing directly on to it. Once satisfied with the display, he may invoke the program generator which automatically produces the code required to produce that screen. It may also produce validation routines stopping only to request what type and range of data is required in a certain field and what error message should appear when the user makes a mistake. These may be produced directly if the Program Generator has access to a Data Dictionary.

- Reusable Code

  Many systems are conceptually similar. Tasks frequently re-appear – menu design, password protection, print routines, date checking, etc. It is possible to build up a library of well-proven, well-documented routines which may be plugged together to make up a system. The *content*, not the logic, will need changing and some patching will be required to make a complete system. However, access to a store of well-tested standard modules should ease development and maintenance.

- Fourth Generation Languages

  However, it is the emergence of Fourth Generation Languages (4GLs) that have largely permitted the adoption of the prototyping approach.

Languages may be seen as passing through three previous generations.

— *First Generation:* Machine Code. Instructing the machine through direct binary code. Closely associated with the architecture of the host processor. Complex to write, read and debug.

— *Second Generation:* Assembler Languages. Uses symbolic codes. Machine instructions given by mnemonic alphabetic codes. Easier to understand although still closely allied to machine architecture.

— *Third Generation:* High Level languages such as Fortran, COBOL, PL/1 and BASIC. Written in procedural code. Largely independent of the hardware architecture permitting portability. Much easier to use.

There is no agreed definition on what constitutes a Fourth Generation Language. Martin (1982) provides a checklist of desirable features for 4GL evaluation, and this is summarised below.

- Centred around a relational database.

- Links to other proprietary databases and other non-database files. This will permit gradual transition to the new development strategy.

- Integrated and active Data Dictionary.

- Simple Query Language. May use a syntactical structure and/or a Query by Forms.

- Integrated Screen Design Tool.

- Dialogue Design Tool. Including generation and manipulation of business graphics.

- Report Generator.

- Procedural Coding facility. This may be, done directly through a conventional language (say COBOL) or indirectly via a code Design Aid (such as an Action Diagrammer).

- Non-Procedural Programming Code.

- Spreadsheets and graphics.

If the 4GL is to provide a complete development facility for the professional programmer, analyst and end user it must clearly have a range of tools to accommodate disparate requirements and skill levels. It must also dovetail with the past development strategy of the organisation and the systems developed under that strategy. 4GLs which require massive re-writes of current operational systems are likely to extend the development backlog, not reduce it.

Prototyping: Summary

Possible applications of prototyping have already been established throughout this book. It is particularly useful in helping users define their requirements and in agreeing the contents, display and technology of the user interface. This issue will be re-visited in the companion text where a formal method of documenting the prototype will be suggested.

However, it must be stressed that prototyping is not a panacea. Two particular disadvantages addressed by the life cycle model (Alavi 1984) are:

- *Management and control:* The absence of definite phases make planning and budgeting very difficult. Documentation and testing also tends to be overlooked.

- Size: It is very difficult to prototype large information systems. Prototyping may lead to inappropriate applications and incompatibility.

10.5 OBJECT ORIENTED ANALYSIS

The exploration of Object-Oriented systems development appears to be coming from two different traditions. The first is from the programming perspective. Programmers using languages such as Smalltalk and Simula sought an approach to program construction which allowed them to use the facilities offered by such languages. They found that the 'top-down' functional decomposition approach developed for conventional languages was inappropriate to this new environment. These Object-Oriented Programming Languages needed a new approach that challenged the established methods and gurus (Cox 1986, Meyer 1988). Meyer in particular has brought the object-oriented approach forward in the life cycle to consider design.

The second tradition is that of information systems development. The construction of generic data models has emerged from both the Yourdon and Information Engineering schools with system templates made available in the commercial marketplace.

This section looks briefly at the emergence of object-oriented development from both these perspectives. The issues and concepts raised will be examined in more detail in the companion text *Introducing Systems Design*.

10.5.1 Generic models

Looking for some similarity between Object-Oriented Programming (OOP) and an emergent Object-Oriented Analysis, Ed Yourdon identified the concept of pattern matching. He found that, in general, OOP designers did not begin applications from scratch but tried to find an existing software pattern that broadly fitted the requirement. The OOP developer tended to assume that a pattern does exist and it is just a matter of finding it. In contrast, Yourdon claims, the information systems community faced with a problem for which they do not have an immediately obvious pattern, immediately start from scratch.

Yourdon refers this pattern approach back to architectural form (Alexander, 1979). However, a similar intellectual approach to problem solution is also evident from Checkland's work briefly discussed in the second chapter of this book. Stage Four of his soft-systems approach builds conceptual models from the CATWOE and then (in Stage 5) compares them with reality. The conceptual models may be examined for

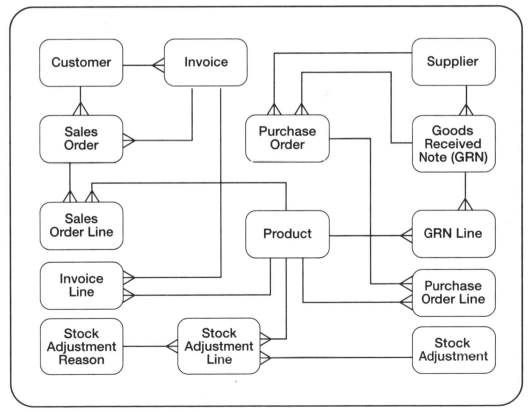

Figure 10.6   Generic data model for a core order processing system

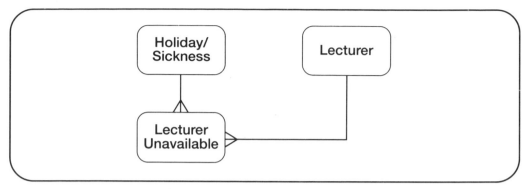

Figure 10.7   Lecturer LDS fragment

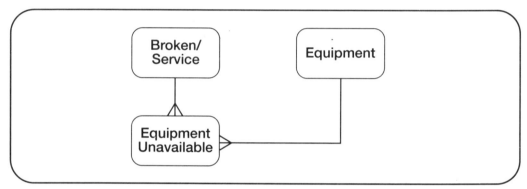

Figure 10.8   Equipment LDS fragment

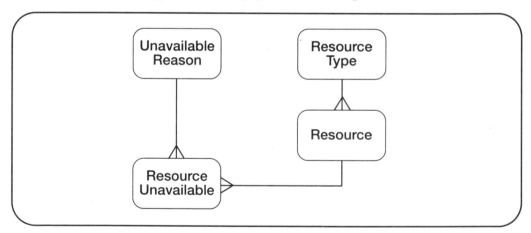

Figure 10.9   Resource LDS fragment

validity "in terms of any other systems thinking which the analyst reveres". It is the point where "other veins may be mined", conceptual models may "be inspected alongside any systems theory which is relevant to human activity systems". The point is that conceptual models will be built to reflect knowledge and insights from elsewhere. Furthermore, it is only after completing the conceptual model that reality is examined and a strategy developed to implement feasible and desirable changes. In this way Checkland effectively develops the required logical model first and then only examines the *current system* to identify the changes needed for the required implementation.

Defining requirements and looking for models that already form a good 'fit' is also evident in Yourdon's application domain models. He feels it should be possible to develop a handbook for designers which gives basic classes of objects that would be expected in a particular application area. We have developed similar concepts in our own company. The Logical Data Structure in Figure 10.6 could form the core of most sales and purchase order processing systems. When we meet a new company requiring a development in this area we compare their requirements with the features provided by this model.

This idea of 'template' solutions has been taken up by software suppliers. JMA Information Engineering (Software Management, November 1991) market template solutions developed with the IEF CASE tool. These templates are essentially system models which include business data requirements, definitions of business activities, design specifications, screen layouts and the logic behind the screens. Users can take these and tailor them to a specific application by changing the logical specification and then producing the code automatically through the CASE tool.

The generalisation of the model can be taken further. The logical data structures of Figures 10.7 and 10.8 were produced from two systems. Actually they are the same system viewed at a more abstract level (*see* Figure 10.9).

10.5.2 Object terminology

Coming from the other direction Meyer asks whether the design of a system should be based around functions (actions or processes) or data (objects). He concludes that function based methods are unsatisfactory and that systems should be constructed around data. Four of his reasons are given below and the first, at least is familiar.

- Functions or processes are more volatile than data. They change to reflect new business requirements and new users. Often this change is incremental and so there is no perceived requirement for major system change. Over time the tasks change dramatically but the classes of data the system manipulates are more persistent.

- The 'top-down' approach is based around functional decomposition. Processes are progressively decomposed until they are sufficently detailed to be implemented in the target programming language. This approach means that the design is based around the most volatile perspective of the system. It should be successful in the short-term, but its long-term effectiveness will be influenced by functional changes. The initial primary functions may quickly become marginal or obsolete to be replaced by processes which have to be accommodated into a design based around outdated assumptions.

- Many process models place too much emphasis on the order in which things happen to data. This often leads to design problems when this order changes. For example, one company known to us recently changed its business sequence from:

— Order, Course-run, Invoice to:

— Order, Invoice, Course-run

Object-oriented design takes a more neutral view of the ordering of processes. The designer defines the operations (or events) and the effect that these have. However, the specification of the order of these events is deferred for as long as possible.

- The 'top-down' approach does not promote re-usability. Elements of the software are developed to satisfy a particular function and hence tend to be "narrowly adapted to the sub-problem that led to their development".

Object-oriented design's (OOD) insistence on the data-centred approach comes as no surprise to practitioners and academics from a data-based background. Martin, through his Information Engineering methodology, has espoused this view for many years. However, OOD challenges a well-documented programming tradition of system development through top-down step-wise refinement.

This data oriented approach leads to clear similarities between the basic constructs of entity and object modelling. For example Meyer uses the term *object* to refer to an entity occurrence and *class* for an entity type. Other writers refer to the class as an object type. Shlaer and Mellor (1988) essentially define entities as objects and present an Information Structure Diagram which is "based on various forms of entity relationship diagram used over the past few years". Another object oriented text outlines an object-relationship diagram which is "essentially the same as an entity-relationship diagram" (Martin and Odell, 1992). The authors also present a further development called an object schema which they compare in depth with the entity-relationship diagram in an appendix to their book.

Classes or object types have particular *features* which include attributes
> *class* EMPLOYEE *feature*
>   employee-no: INTEGER;
>   employee-name: STRING;
>   department: STRING
> *end* — class EMPLOYEE

An attribute is a component of the class that will give a field in each object of that class.

References between classes are declared as follows:
> *class* DEPARTMENT *feature*
>   department-code: INTEGER;
>   department-name: STRING;
>   department-head: STRING
> *end* — class DEPARTMENT

> *class* EMPLOYEE *feature*
>   employee-no: INTEGER;
>   employee-name: STRING;
>   department: DEPARTMENT
> *end* — class EMPLOYEE

This is analogous to the foreign key concept introduced in Chapter 6. In Meyer's terminology EMPLOYEE is said to be a *client* of DEPARTMENT and DEPARTMENT a *supplier* of EMPLOYEE.

The reference is assumed to be void unless a new object of the supplier class type is created and associated with the reference.

Shlaer and Mellor refer to the foreign key as a a referential attribute.

However, the distinction of OOD lies in its desire to include the process specification within the data specification and hence ensure that this functional perspective is not neglected. Thus classes are also defined in terms of the operations that affect instances of a class. Two kinds of routine are suggested; procedures and functions.

A procedure changes the state of an object where the state is defined as the current values of fields. A procedure call changes the values of one or more fields and hence is said to change the state of the object. For example for the class EMPLOYEE a procedure will be defined for changing the grade and pay-rate.

class EMPLOYEE *feature*
    employee-no: INTEGER;
    employee-name: STRING;
    working-week: INTEGER;
    pay-rate: INTEGER;
    grade: CHARACTER;
    salary: INTEGER;
    department: DEPARTMENT
end — class EMPLOYEE

In contrast, a function computes a value deduced from the state of the object. In the above example the field salary is calculated from pay-rate multiplied by working-weeks.

Pre-conditions and post-conditions may also be specified for routines. The pre-conditions express the properties that must hold when the routine is called. Consider the following;

class PAYMENT *feature*
    payment-voucher: INTEGER;
    supplier: SUPPLIER1;
    payment-date: INTEGER;
    payment-amnt: INTEGER;
    authorised: BOOLEAN;
    paid-indicator: BOOLEAN
end — class PAYMENT

The procedure 'pay supplier' will only be allowed if the field authorised is equal to Y(es).

If the pre-condition is met then the procedure runs and eventually sets the post-conditions;

    payment-date = todays-date
    paid-indicator = Y(es)

Thus the post-condition describes the properties that the routine guarantees when it returns.

Permanent guarantees can also be specified

Payment-amnt <= Invoice-amnt

Finally, a class may be designed as an extension or specialisation of another class. This is the notion of inheritance. Inheritance may define classes which may be a sub-set or super-set of another class. For example the class ORDER may have a super-set called STANDING ORDER and a sub-set termed PART ORDER. The range of this interpretation leads to the general concept of heir and descendent. The class ORDER in our example is inherited by both STANDING ORDER and PART ORDER. The implication of this is that all features available to ORDER become automatically available to STANDING ORDER and PART ORDER. Further features can be added to the inheritors for their own specific purposes.

Clearly the concept of inheritance raises the issue of when to close a class. The choice is between closing the class and adding new classes or extending the class to include new features, but hiding those features from particular clients. The latter is termed information hiding where certain features are only exported to defined classes.

Most structured methodologies have gradually moved towards this function within data perspective with the process model (usually represented by the DFD) being downgraded. In the context of the models introduced in this book the Entity Life History (ELH) is constructed for each entity (or object) on the Logical Data Structure. It shows how events (not processes) update the static data structure. However, this approach involves the use of two models (the ELH and the LDS) which are often presented separately. It might be more effective if the ELH was seen as subservient to the data model as after all, the effects of events are documented within the context of each entity or object. The ELH also has the advantage of being a graphical model which is usually easily understood by users. This contrasts with the (usually stylised) textual specifications often adopted by OOD class definitions.

Martin and Odell also examine different event types in their consideration of object behaviour and describe the following kinds of state changes, all of which should be familiar from our consideration of entity life histories.

— Events that create an object.

— Events that terminate an object.

— Events that classify an object as an instance of an object type. For example, an employee becomes a manager.

— Events that declassify an object.

— Events that change an object from one classification to another.

— Events that change attributes within the object.

Martin and Odell have also developed a graphical model for describing the detailed life cycle of an event. This is an event schema (*see* Figure 10.10) with high level activities described in an object-flow diagram, a model similar to the traditional DFD. The state transition diagram is also presented as a possible way of describing the activity in Martin and Odell's recommended diagramming standards.

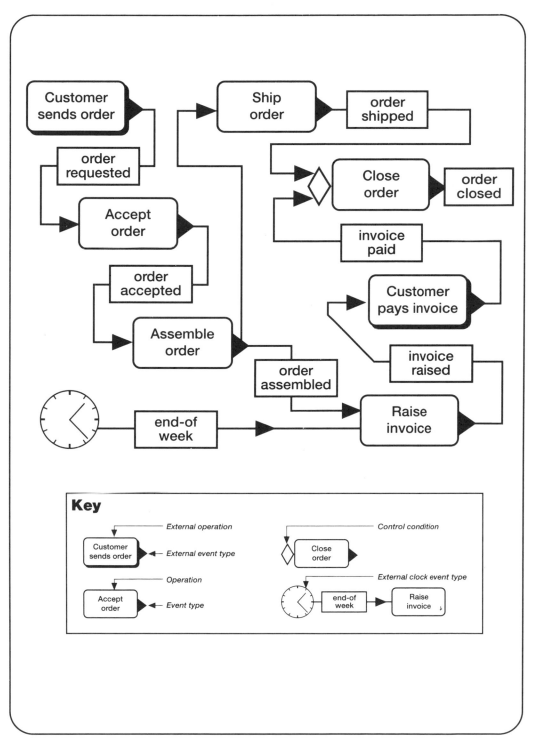

Figure 10.10   Event Schema

10.6 SUMMARY

This chapter has introduced four further system perspectives.

- The extended real-time data flow diagramming notation of Ward and Mellor. Process specification using state transition diagrams will be considered in the companion text *Introducing Systems Design*.

- An alternate data modelling convention proposed by Chen and extended by Howe. This notation is often used by American CASE tools.

- The principles of prototyping. Documenting and developing a prototype for constructing the user interface is discussed in detail in *Introducing Systems Design*.

- The emerging concepts of Object Oriented development. Object Oriented Design (OOD) is discussed in more detail in *Introducing Systems Design*.

11 Systems Design

This chapter looks briefly ahead to the complementary text – Introducing Systems Design.

So far, inputs and outputs have been defined in terms of their logical data contents and documented in the data dictionary. Physical input and output design will allocate technologies and layouts to these data flows and consider the dialogues required to enter and retrieve data. The role of prototyping the interface design will be of particular concern and documenting and controlling the prototype will be addressed.

Physical process design will be examined from three perspectives:

- Top-down functional decomposition using action diagrams.

- Object oriented design.

- Real-time design.

Each of these gives a different perspective and the relative merits of each will be highlighted.

Physical data design is primarily concerned with flexing a normalised data model to produce an optimised design. This book has introduced normalisation in a relatively intuitive way. The companion text defines a more mechanistic procedure for identifying relations, showing the progression through first, second and third normal forms. However, it is also recognised that implementing normalised data files directly might lead to poor performance and so strategies for flexing the models are introduced.

Finally, controls and implementation issues and methods are discussed, including the legal requirements of the Data Protection and Computer Misuse Act.

Appendix A
Case Study

1: Introduction

2: Standards

3: Reporting requirements

4: Volumes

5: Appendix 1

1.   INTRODUCTION

A   Woodland Software Services provides software support and consultancy to the IBM mainframe market. It specialises in turnkey systems and training, particularly in manufacturing and distribution.

B   It currently has 140 staff and a turnover of £4.5 million per year. All staff are based at the company headquarters in Edinburgh. Projects are undertaken throughout the United Kingdom and Europe. The Managing Director, Mike Smith, founded the company in 1973.

C   A year ago a rudimentary project recording system was established to monitor project effort and expenditure. This reflected concern amongst directors about the profitability of certain projects and the lack of overall control of resources.

D   The project reporting system was established using simple manual forms and procedures. This was perceived as a trial to see whether it was worthwhile establishing a more comprehensive computer system.

E   The current manual system is maintained by Mike Smith's personal assistant Rosemary Geddes. A discussion record of an interview with Miss Geddes is attached to this report as Appendix 1.

F   Continuing disquiet about project performance and the recent loss of a major customer has convinced Mike Smith of the necessity to proceed with computerisation.

G   Woodland Software Services currently has two minicomputers and 32 networked microcomputers. One of the minicomputers is reserved for administrative services and currently supports the company's accounting and payroll functions. This machine is lightly loaded and is the likely target machine for the proposed project monitoring system. This minicomputer already has established gateways to the microcomputer network. Woodland specialise in 4GL and database development and so the proposed software development will probably be undertaken in DB/1, the firm's usual development tool.

2. STANDARDS

A   The standards system required by Mike Smith is designed to extend his current manual arrangements.

Each project will have a unique name and consist of a series of project phases. For example, the Terrafimal project will pass through the following phases:

– Business Survey
– Feasibility Study
– Systems Analysis
– Systems Design
– Program Specification
– Programming
– System testing
– Acceptance testing
– Implementation

The current target date of the project is defined as the end of the Feasibility Study as there is no certainty that the project will progress beyond this stage. Project phases for each project may be different but they should always fit in with the general phases defined above. This phasing and naming arrangement is supported by the current manual system.

B   However, the required standards system will extend this to project tasks. Tasks are currently recorded on the manual system but there is no agreement of naming conventions. There is also no cross-reference between the same tasks on different projects. The new system should introduce standardisation as well as permitting task analysis. A preliminary consideration of tasks has revealed the following:

Phase	No. of Tasks
Business Survey	8
Feasibility Study	12
Systems Analysis	25
Systems Design	20
Program Specification	35
Programming	15
System testing	7
Acceptance testing	12
Implementation	8

Each project must be defined in terms of the standard phases and tasks. However, new phases and tasks may be added to the standards system by agreement with Mike Smith.

C The main objective of the proposed standards system is control. However, the problems of project cost estimation are also recognised. Mike Smith would like to see each task defined as a number of standard hours that can be used as a basis for project costing. It is recognised that the same tasks may have different weightings on different projects. However, some sort of standard for tasks (particularly if it is kept under review) would provide "ballpark" figures for project estimation.

3. REPORTING REQUIREMENTS

Report 1

Project Analysis

Report requested by project. Available as a screen and print option. Estimate and target finish date remain fixed throughout the project. Available for all live and completed projects.

Project: Taylor-Goodall				Date: 06/04/89	
Phase	Estimate (hours)	To date	To finish	Target date	Complete date
Business survey	80	56	0	31-01-89	24-01-89
Feasibility Study	200	160	0	06-03-89	01-03-89
Systems Analysis	320	80	240	17-05-89	
Systems Design	240	0	240	25-05-89	
Program Specification	240	0	240	13-06-89	
Programming	840	0	840	20-08-89	
System testing1	20	0	120	20-09-89	
Acceptance testing	40	0	40	25-09-89	
Implementation	80	0	80	31-09-89	
Total	2160	296	1800		

Report 2

Phase Analysis

Phase analysis is required by project and across projects, for example:

Phase Analysis: Business Survey			Date: 06/04/89	
Project	Estimate (hrs)	Actual (hrs)	Difference (hrs)	Complete date
Taylor-Goodall	80	56	24	24-01-89
Terrafimal	110	146	(36)	28-01-89
Hughes Lifts	140	30		
Arad Systems	230	10		
Average	95	101	(6)	
Cum. Average	70	90	(20)	

The average figure gives the average for all completed projects on the report. The Cum. Average (cumulative average) gives the average for all projects completed in the last three years.

It is envisaged that a similar report will be available for task analysis across projects.

Report 3

Employee Analysis

A time sheet analysis produced at the end of each week. This should be available by employee (an example is given below), by project, by employee within project and by company.

Time Sheet Analysis: Employee Employee Name: Steve Jones		Week Ending: 14/04/89	
	Normal Time	Normal Overtime	Weekend Overtime
Saturday			
Sunday			
Monday	7.5		
Tuesday	7.5	2.0	
Wednesday	7.5		
Thursday	7.5	2.5	
Friday	7.5	2.0	
Total	37.5	6.5	

Time sheet detail will be automatically passed to the current computerised payroll system.

Report 4

Expense Analysis

The expense analysis should be available by employee, by project and by company. The budgeted estimate is given on the project report only (example given below). This is available to project managers.

Project Cost Analysis: Taylor-Goodall				Date:06/04/89	
	Staff Costs	Non-claimable Expenses	Budget	Cumulative Actual	Budget
Business Survey	2520	450	4800	2970	4800
Feasibility Study	7200	750	12000	10930	16800
Systems Analysis	3600	240	19200		
Systems Design			14400		
Program Specification			14400		
Programming			33600		
System testing			4800		
Acceptance testing			1600		
Implementation			4800		
Total			109600		

Cumulative figures are only shown for completed phases.

The expense analysis for employees is produced at the end of each week and shows:

    Project Name
    Project Phase
    Expense Type: Travelling
                 Subsistence
                 Accommodation
                 Miscellaneous
    Expense Amount : for each expense type.

Both expense type and expense amount are further sub-divided by the accounts section into claimable and non-claimable. Claimable expenses are invoiced to the client.

Further Reports

Report 6 will provide an analysis of estimated task length and the average actual task length from projects that have completed these tasks in the last three years (see later). This information will be used to vary standard estimates.

4. VOLUMES

A The number of projects for the last three years is given below:

1988	65
1987	45
1986	43

The projected number of projects for the next three years is:

1989	90
1990	110
1991	120

B A phase breakdown for projects on 1988 is shown below:

Phase	Number of Projects
Business Survey	45
Feasibility Study	54
Systems Analysis	59
Systems Design	60
Program Specification	55
Programming	53
System testing	49
Acceptance testing	52
Implementation	58

C Projected staffing is shown below

1989	140
1990	175
1991	192

D Projects are considered to be 'live' if they have been completed within the last three complete calendar years. Projects completed before this are archived and do not contribute to the overall statistics. All live projects are required to be viewed on-line.

5. APPENDIX 1: DISCUSSION RECORD WITH ROSEMARY GEDDES

DISCUSSION RECORD		Sheet *1*
Description *Project recording*	**System** *TIME-SYS*	of *4*
PARTICIPANTS *Rosemary Geddes (RG)* *Steven Jones*	**Date:** *11/03/89* **Location:** *HQ* **Duration:** *2 ½ hrs*	
OBJECTIVE: *Current time recording system*		

SUMMARY:

i/ Mike Smith notifies RG of a new project heading and the standard phases that will take place in the project.

For example: Allied-Data Systems (project name)

Feasibility Study
Systems Analysis } *(project phases)*
Systems Design

RG raises a project record sheet for each phase.

ii/ Each week RG receives timesheets from staff. She photocopies this sheet and sends the original to the payroll clerk. The photocopy is used to update the relevant project record sheet. After updating the timesheet is filed in an employee file. This is an envelope style folder with one folder for each employee. They are kept in alphabetical sequence in a filing cabinet drawer.

Discussion Record with Rosemary Geddes

DISCUSSION RECORD		Sheet *2*
Description *Project recording*	System *TIME-SYS*	of *4*
PARTICIPANTS *Rosemary Geddes (RG)* *Steven Jones*	Date: *11/03/89* Location: *HQ* Duration: *2 1/2 hrs*	

OBJECTIVE: *Current time recording system*

SUMMARY:

iii/ Many staff are unaware of the correct naming of phases and so these have to be checked back with the employee who filled in the form.

iv/ There may be many project record sheets for each project phase and sub-totals are carried forward.

v/ There is no current use of project tasks because the naming conventions vary so much. However, the tasks are noted on the project record RG sees no point in this...

vi/ At the end of each week RG produces a project summary report which gives
Project name;
Cost of project to date;
Expenditure on project in last week;

Discussion Record with Rosemary Geddes

DISCUSSION RECORD		Sheet *3*
Description *Project recording*	System *TIME-SYS*	of *4*
PARTICIPANTS *Rosemary Geddes (RG)* *Steven Jones*	Date: *11/03/89* Location: *HQ* Duration: *2 ½ hrs*	

OBJECTIVE: *Current time recording system*

SUMMARY:

RG produces these from summary totals maintained in an exercise book. She finds this very time consuming although she is helped by a YTS employee.
The exercise book holds the following information

> *Project name: Alllied-Data Systems*
> *Cost of project to date: £17,100*
> *Cost of project, week ending 20/01/89: £2171.00*

The cost of project to date includes the week-ending cost.

vii/ Project costing includes expenses. RG receives a copy of this from accounts. There is no current analysis by expense claim type.

viii/ RG not sure why figures are kept in such depth. Project managers seem to give them little credibility and often question her arithmetic.

Discussion Record with Rosemary Geddes

DISCUSSION RECORD		Sheet *4*
Description *Project recording*	System *TIME-SYS*	of *4*
PARTICIPANTS *Rosemary Geddes (RG)* *Steven Jones*	Date: *11/03/89* Location: *HQ* Duration: *2 1/2 hrs*	

OBJECTIVE: *Current time recording system*

SUMMARY:

ix/ *At the end of the project, the project record sheets*
are placed in a box file in the company vaults. A
summary sheet is sent to project manager detailing
the total cost of each phase.

x/ *RG finds the job particularly time consuming. Mike*
Smith is making an increasing number of ad hoc
requests which demand time consuming searchings,
retrieval and calculation. Example given:

> *Look at the time taken on software evaluation*
> *in the last five projects.*

xi/ *RG would like to see*

- an easier way of maintaining the figures

- a clearer idea of what they are used for.

Discussion Record with Rosemary Geddes

Project Name		Project Phase				Manager		
TERRAFIMAL		*FEASIBILITY STUDY*				*FRANK JONES*		
Target Date		**Target Date**				**Location**		
30/01/89		*30/01/89*				*BOURNEMOUTH*		

Project Task	Empl. Inits	Date	Time Spent N	O	Claim	Type	Other
Terms of reference	*FJ*	*11/12/88*	*7½*	*2*	*14.10*	*Petrol*	
"	*FJ*	*14/12/88*	*4*	*–*	*21.00*	*B.R.*	
Systems investigation	*MJ*	*02/01/89*	*7½*	*–*	*27.30*	*Petrol*	
					4.50	*Lunch*	
"	*MJ*	*03/01/89*	*7½*	*–*	*45.00*	*Hotel*	
					7.50	*Misc.*	
"	*MJ*	*04/01/89*	*7½*	*3*	*–*	*–*	
"	*MJ*	*05/01/89*	*7½*	*–*	*45.00*	*Hotel*	
					7.50	*Misc.*	
"	*MJ*	*06/01/89*	*7½*	*3*	*12.50*	*Lunch*	
Project meeting	*FJ*	*10/01/89*	*7½*	*2*	*32.50*	*Petrol*	
"	*MJ*	*10/01/89*	*7½*	*2*			
"	*FP*	*10/01/89*	*7½*	*2*			
Hardware evaluation	*MJ*	*12/01/89*	*4*	*–*			
Software evaluation	*MJ*	*13/01/89*	*7*	*–*			
Hardware evaluation	*MJ*	*16/01/89*	*3*	*–*	*27.30*	*Travelling*	
				–	*3.50*	*Parking*	
"	*MJ*	*17/01/89*	*4*	*–*	*4.50*	*Lunch*	
Walkthrough	*FJ*	*23/01/89*	*7*	*–*			
	MJ		*7*	*–*			
Report writing	*MJ*	*24/01/89*	*7½*	*–*			
	MJ	*25/01/89*	*7½*	*–*			
	MJ	*26/01/89*	*7½*				
		Carried Forward	*126*	*14*	*252.20*		

Case Study

WOODLAND SOFTWARE SERVICES			Employee Name: _Mike Jackson_		W/E _20/01/89_
Project Name	Project Phase	Project Task	Date	Expense Claim	Expense Type	Other
Terrafirma	Feasibility Study	Hardware Evaluation	17/01/89	£27.30 3.50 4.50 £35.30	Travelling Parking Lunch	
Neville-Fraser	Preliminary Meeting	–	20/01/89	£92.50 6.50 1.75 £100.75	Travelling Subsistence Parking	
						Mike Jackson

Case Study

WOODLAND SOFTWARE SERVICES

Employee Name: *Mike Jackson* **W/E** *20/01/89*

Project Name	Project Phase	Project Task	Date	Time Spent Normal	Time Spent Overtime	Other
Compustar Terrafimal	System Testing Feasibility Study	Module Test Hardware Evaluation	16/01/89 "	4 3		
N/A Compustar Terrafimal	Admin System Testing Feasibility Study	Module Test Hardware evaluation report	17/01/89 "	1/2 3 1/2 4		
		18/01/ ill				
Neville-Fraser "	Preliminary meeting "	– Project bid	19/01/89 20/01/89 TOTAL Sick	7 1/2 7 1/2 30 7 1/2	2 1 3	*Janice*

Mike Jackson

Appendix B
Bibliography

CHAPTER 1

De Marco T. *Structured Analysis and System Specification* Prentice Hall, 1979
Benyon D. and Skidmore S. *Towards a tool kit for the Systems Analyst* Computer
    Journal, 30, 1.

CHAPTER 2

Ackoff R. *A Concept of Corporate Planning* Wiley Interscience, 1970
Ansoff I. *Corporate Strategy* Penguin Business, 1987
Checkland P. *Systems Thinking, Systems Practice* Wiley, 1981
Hussey D. *Corporate Planning - Theory and Practice* Pergamon, 1982
Grindley K. and Humble G. *The Effective Computer* McGraw-Hill, 1973
Mintzberg H. *Mintzberg on Management* Frce Press, 1989
Wilson B. *Systems: Concepts, Methodologies and Applications* Wiley, 1984
Wood-Harper A, Antill L, Avison D. *Information Systems Definition: The Multiview
    Approach* Blackwell, 1985

CHAPTER 3

Chapin N. *Economic Evaluation in Systems Analysis and Design - a Foundation for
    the 1980s.* Cotterman W *et al*, eds, North-Holland, 1981
Collins G., Blay G. *Structured Systems Development Techniques: Strategic Planning
    to System Testing*, Pitman, 1982
Lumby S. *Investment Appraisal*, Nelson, 1981
Parkin A. *Systems Analysis*, Edward Arnold, 1980

CHAPTER 4

Autostate. *Communications of the ACM*, July 1964, vol 7, no 7, pp 425-432
Cherry C. *On Human Communication*, MIT Press, 1978
Daniels A. Yeates D. *Basic Training in Systems Analysis*, Pitman, 1971

Groner C. Hopwood M. Palley N. Sibley W. Requirements Analysis in Clinical Research Information Processing - A Case Study, *Computer*. September 1979, vol 12, no 9, pp 100-108

Hein K. Information System Model and Architecture Generator, *IBM Systems Journal*, 1985, 24, 3/4, pp 213-215

Kimmerly W. Restricted Vision, *Datamation*, 6 March 1984

McMenamin S. Palmer J. *Essential Systems Analysis*, Yourdon Press 1984

Moser S. Kalton G. *Survey Methods in Social Investigation*, Heinemann 1971

Parkin A. Thornton S. Holley P. Can Fact-finding Be Automated *Automating Systems Development*, Benyon D, Skidmore S, Plenum Press, 1987

Rosenteel G. Why Systems Analysis Training Fails. *Computer World* November 1987

Sumner S. and Sitek J. *Are Structured Methods for Systems Analysis Being Used?* Journal of Systems Management June 1986, pp 18-23

Yourdon E. *Whatever Happened to Structured Analysis?* Datamation, 1 June 1986

CHAPTER 5

Gane C. and Sarson T. *Structured Systems Analysis* Improved System Technologies, 1980

Yourdon E. *Modern Structured Analysis* Prentice Hall, 1989

CHAPTER 6

Benyon D. *Information and Data Modelling* Blackwell Scientific Publications, 1990

Chen P. *The Entity-relationship Model: Towards a Unified View of Data* ACM Transactions on Database Systems, March 1976, pp 9-36

Howe D. *Data Analysis for Data Base Design* Edward Arnold, 1983

Martin J and McClure C. *Diagramming Techniques for Analysts and Programmers* Prentice Hall, 1985

CHAPTER 7

Ward P. and Mellor S. *Structured Development for Real-Time Systems* Yourdon, 1985

Fergus R.M. Decision Tables – What, Why and How, *Proceedings College and University Machine Records Conference*, University of Michigan, 1969, pp 1–20

Lew A. Proof of Correctness of Decision Table Programs, *Computer Journal*, 1984, vol 27, No 3, pp 230–232

CHAPTER 8

Skidmore S. Farmer R. and Mills G. *SSADM-Models and Methods* NCC-Blackwell, 1992

CHAPTER 9

British Computer Society: *Data Dictionary Systems Working Party* BCS, 1977

Leong-Hong B. Plagman B. *Data Dictionary/Directory Systems: Administration, Implementation and Usage,* Wiley, 1982

Mayne A. *Data Dictionary Systems: A Technical Review,* NCC Publications, 1984

Plagman B. Moss C. *Alternative Architecture for Active Data Dictionary/Directory Systems,* Auerbach, 1978

CHAPTER 10

Alavi M. *An Assessment of the Prototyping Approach to Information Systems Development* Communications of the ACM, June 1984, vol 27 No 6

Boehm B *Prototyping versus Specifying: A Multiproject Experiment* IEEE Transactions on Software Engineering, May 1984, vol SE-10 no 3

Cox B. *Object-Oriented Programming: An Evolutionary Approach* Addison Wesley, 1986

Coad P. and Yourdon E. *Object-oriented analysis* Yourdon Press, 1991

Martin J. and Odell J. *Object-oriented analysis and design* Prentice Hall, 1992

Meyer B. *Object-oriented Software Construction* Prentice Hall, 1988

Martin J. *Application Development without programmers* Prentice Hall, 1982

Shlaer S. and Mellor S. *Object-oriented systems analysis* Yourdon Press, 1988

GENERAL

Date C. *An introduction to Database Systems vol 1, 4th edition Addison Wesley, 1986*

Kent W. *A simple guide to Five Normal Forms in Relational Database Theory* Communications of the ACM, February 1983, pp 120-125

Jackson M *Systems development,* Prentice Hall, 1983

Connor D. *Information System Specification and Design Road Map* Prentice Hall, 1985

NCC *SSADM version 4 manual* NCC Blackwell, 1990

Appendix C
Exercises

CHAPTER 2

Q1 Read the Woodland Software Services study. Draw a Rich Picture of the systems described in the narrative and the interview with Rosemary Geddes.

Q2 Suggest a root definition for the system from the perspective of the Managing Director.

Q3 Draw a conceptual model of the system based on the agreed root definition.

Q4 What strategic objectives might the required system contribute to? Discuss the organisational benefit of such a system.

CHAPTER 3

Q1 What technical and operational feasibility issues are likely to have to be considered in the Woodland Software Services case study?

Q2 Woodland Software Services are currently looking to replace part of their computer network. Two suppliers have been asked to tender for the project. The relevant costs of each project is given below.

Year	Supplier A		Supplier B	
	Cost	Savings	Cost	Savings
0	100,000	0	75,000	0
1	0	20,000	25,000	0
2	0	30,000	0	60,000
3	0	30,000	0	40,000
4	0	40,000	0	40,000
5	10,000	50,000	10,000	40,000
	110,000	170,000	110,000	180,000

1) If the current interest rate is 10%, which supplier should be given the contract using the Net Present Value evaluation Method.

2) What other factors should be taken into account when choosing a supplier?

3) If the interest rate fell to 5% would this change your decision?

Q3 There are two main approaches to project appraisal using discounted cash flow. These are Net Present Value (NPV) and Internal Rate of Return (IRR). Investigate these approaches and present a short tutorial paper that shows the use of these two methods in a hypothetical example.

Discount	10%
Year	
0	1.0000
1	0.9091
2	0.8264
3	0.7513
4	0.6830
5	0.6290

CHAPTER 4

Q1 The following extract is from a meeting between an analyst and an order clerk.

Analyst: "Hello Glenys, I want to concentrate today on looking at the procedures you undertake to process the orders that you receive directly from the customer, is that OK?"

Order Clerk: "Yes, I suppose so, will it take long?"

Analyst: "Only a few minutes . So can you tell me how you handle a typical order?"

Order Clerk: "Well, I first check the customer for credit-worthiness. I do this by looking at the relevant customer folder where the customer's current credit value and credit limit are held."

Analyst: "The current credit value being the balance they currently owe you and the credit limit the maximum balance allowed?"

Order Clerk: "Exactly, I must make sure that the value of the order added to the current credit value doesn't take them over their limit."

Analyst: "What happens if it does?"

Order Clerk: "I make a copy of the order and place it in the pending tray. I then send the original order to accounts and wait to see if I am able to process the order."

Analyst: "When do they tell you if you can process the order?"

Order Clerk: "Within 24 hours. When I receive the original order back from accounts it is marked with an accept or reject label and signed by a credit clerk. For failed orders I raise a standard reject letter and send it, together with the order, back to the customer. I destroy the copy that I put in the pending tray."

Analyst: "And for accepted orders?"

Order Clerk: "I process these normally."

Analyst: "What about the copy in the pending tray?"

Order Clerk: "Oh, I just throw that away."

Analyst: "What is the normal process for an order after it has been checked?"

Order Clerk: "I raise a standard order confirmation and send a copy to the customer. A further copy goes to despatch and they use that for packing the order.

	I keep the blue copy here with the customer's original order form."
Analyst:	"What do you use that for?"
Order Clerk:	"What for?"
Analyst:	"The blue copy?"
Order Clerk:	"Well, for answering queries... and of course Mavis uses it in the end of day update."
Analyst:	"The end of day update?"
Order Clerk:	Yes, she updates the current credit value for each customer at the end of the day. Didn't she tell you that?"
Analyst:	"Mavis?... this is Miss Clark, the Order Supervisor?"
Order Clerk:	"Yes."
Analyst:	"Perhaps she did, I'll check the notes. One last thing. Then you calculate the order value to see if it takes the current credit value over the credit limit."
Order Clerk:	"Yes."
Analyst:	"Where do you get the price of each product from, because it doesn't appear to be on the order form?"
Order Clerk:	"From the Price List of course. Look I've got one here. Didn't Mavis mention that either?"

Produce a discussion Record of this meeting.

What questions would you wish to ask at the next interview?

Q2 Undertake Document Analysis of the forms used by Rosemary Geddes in The Woodland Software Services case study.

What questions would you wish to ask at the next interview with Rosemary Geddes?

CHAPTER 5

Q1 The following data was obtained by interviewing the personnel in the departments concerned.

The design draughtsman creates a design drawing. When finished he removes from file the Part Number Register and examines it for the last used part number. The draughtsman enters the next part number and part description in the register and the part number on the drawing and refiles the Part Number Register.

The design draughtsman passes the design drawing to his Section Leader who checks for any errors. If he finds an error he marks it in red pencil and returns it to the draughtsman, who makes the necessary correction and resubmits the drawing to the Section Leader. The Section Leader rechecks the drawing and passes it to the Chief Designer for signature. If the drawing is correct when first checked the Section Leader passes it straight to the Chief Designer. The Chief Designer signs the drawing in the space provided and passes it to the Print Room.

The Print Room prints two copies of the drawing on a dyeline printer and, placing these together, passes them to the Design Office. The original design drawing is returned separately to the Design Office wher it is filed in the 'Master ' file.

On reaching the Design Office the No.1 print is stamped 'Advance Issue' by the draughtsman who then separates the copies, files the No. 2 copy in the Design

Office and passes the other to the Technical Engineer. The Technical Engineer checks to see if the drawing is the last for the assembly, if not he files it in a 'Pending' file. If it is the last he withdraws from a 'Pending' file the remaining prints for the assembly, places them together and creates a Parts List (handwritten).

The Technical Engineer, having raised the Parts List, sends it to the machine room supervisor who passes it to the punch operator. The punch operator punches a Breakdown Pack from the Parts List, then machine sorts the pack into part number order within assembly level. The Parts List is returned to the Technical Engineer (Design Office) who files it.

Having sorted the cards the punch operator uses a tabulator to produce a complete Breakdown Parts List. The cards are filed in the machine room and the Parts List passed to the Technical Engineer who then withdraws a complete set of No. 2 drawing prints from the file, places the whole lot together and passes them to the Process Planner.

The personnel working within the Design Office are:

Draughtsmen
Section Leaders
Chief Designer
Technical Engineers

Required; Draw a system flowchart of these procedures.

Q2 The following extract is from a meeting between an analyst and an order clerk. (See Q1, Chapter 4).

Draw a data flow diagram to model the system described by the clerk.

Review the question list you identified in the previous chapter and add further questions you would want answered at the next interview.

Q3 Draw a current system data flow diagram for the Woodland Software Services case study. Base this diagram on the interview with Rosemary Geddes. There is a vital piece of information that she fails to provide in this interview which she would need to produce the figures she maintains.

Q4 Review different data flow diagramming conventions and present a comparison.

CHAPTER 6

Q1 In a library administration system are the following items likely to be entities (entity types), attributes, an occurrence of an entity or none of these.

Book
ISBN (International Standard Book Number)
Author
Author-name
Librarian
Art Garfunkel
The Wind in the Willows
Loan-date
Fine

Q2 A farmer wishes to keep computerised records on the milk and calf production of his dairy herd. All calves produced are sold and not added to the dairy herd. Each cow has a name and a date of birth and will produce milk for a lactation period after the birth of a calf or calves. The milk recordings for each cow in terms of litres are taken each day. The information required for each pregnancy of a cow are the bull's name, date of mating, date of birth, and each calf's weight at birth and sex. The system is to provide the following information to the farmer:

(a) The details of all births of calves attributed to each bull.

(b) The milk yield of a cow over a particular lactation period.

Draw a logical data structure for the system and suggest the tables which will support the farmer's requirements for information.

Q3 Normalise the following table.

Examination invigilation schedule

Room	Date	Time	Course No.	Title	Examination	Students	Invigilator	Staff Room
W5.13	11/06/87	14.00	125	Comp Sci	Programming	80	Skidmore	W2.17
W6.11	12/06/87	14.00	125	Comp Sci	Systems	80	Mills	W8.24
B1.23	12/06/87	09.00	207	Bus Stud	Modelling	25	Smith	W5.7
W5.13	12/06/87	14.00	402	Maths	Lin Algebra	32	Henton	B6.14
W6.11	13/06/87	14.00	207	Bus Stud	Systems	25	Mills	W8.24
W5.13	13/06/87	14.00	125	Comp Sci	Networks	80	Henton	B6.14
B1.23	14/06/87	09.00	207	Bus Stud	Programming	25	Skidmore	W2.17

Q4 Construct a current Logical Data Structure for the Woodland Software Services System. Base this diagram on the narrative of the case study, the interview with Rosemary Geddes, *and particularly*, on the forms currently used in the manual system. construct normalised tables to support the diagram and fully name all relationships. Document each table in an entity description form.

CHAPTER 7

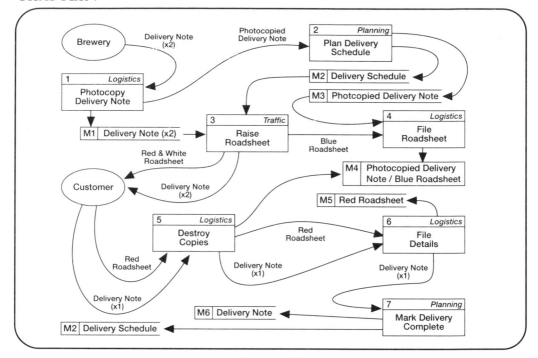

Figure C1

Q1 Convert the current physical data flow diagram above into its logical equivalent.

Q2 Produce a current logical DFD for Woodland Software Services.

Q3 Construct a Problem/Requirements list for Woodland Software Services. Base this on the narrative of the case study and Rosemary Geddes' interview.

Q4 Extend the current logical DFD into a required logical DFD(s) to fulfil all the requirements identified in the previous question.

Extend the logical data structure produced earlier to reflect new requirements. Update the tables and ensure that they are in Third Normal Form.

Q5 Fabrics Galore Ltd has a standard method of receiving yarn stock requests. As requests are phoned through from the factories, a sales clerk checks to see if any stock is held. If no stock is available then a phone call is made to the supplier requesting a special delivery and the stock request is filed in the 'HOLD' file. If stock is available, then the clerk checks to see if the stock meets the requirements of the works order. If stocks do not meet the works order requirements, the clerk checks to see if there is a purchase order outstanding on the yarn stock. If such a purchase order has already been made then the clerk sends a postcard requesting urgent delivery. In both cases, the available stock is sent, although it comprises only part of the order, and the balance is adjusted accordingly. The stock request is then filed in the 'HOLD' file so that the rest of the order can be fulfilled when yarn stocks are replenished.

However, if the yarn stocks meet order requirements, the sales clerk examines the stock to see if they are below reorder level. If they are below this level then the clerk must check to see if a purchase order is outstanding. If there is an order outstanding then a postcard is sent demanding urgent delivery, but if no order has been made then a purchase order form is made out and despatched to the yarn supplier. If the balance is not below the reorder level then these actions are not required. In all cases when the stock is suffcient to meet orders the whole batch of the order is despatched and the stock balance is adjusted accordingly.

(a) Draw a decision table to record the above procedures.

(b) Suggest alternative methods of representing the above rules.

Q6 Discuss the relative merits of system flow charts, dataflow diagrams, decision tables and decision trees. which do you find easiest to draw and understand.

Q7 Construct the Structured English description of the first process on the required logical DFD of the Woodland Software Services system.

CHAPTER 8

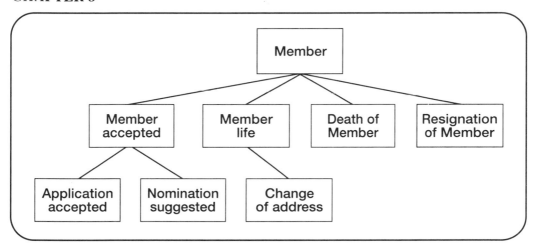

Figure C2

Q1 Identify the mistakes made in the above Entity Life History showing the Member entity of a Golf Club system

Member (*Member-no*, member-name, member-address, date applied/nominated, nominated-by, date of death/resignation

Redraw the ELH using any reasonable assumptions. State your assumptions

Q2 Produce the entity/event matrix for the Woodland Software Services system.

Q3 Construct the Entity Life Histories for the Woodland Software Services system.

Q4 Add state indicators to the ELHs of the Woodland Software Services system.

CHAPTER 9

Q1 Construct the Entity/Data Store grid for the Woodland Software Services system.

Q2 Construct representative Data Dictionary entries for the Woodland Software Services system.

Q3 One of the problems of examining data dictionary software is the wide range of facilities and nomenclature used by the vendors. By using a suitable reference

book (such as Mayne):
- Define what six different packages call the concept of the meta-entity.
- Access the activity of six different packages.

Q4 Dicuss the problems of introducing a data dictionary system into an organisation.

Q5 Write the English narrative that describes the rules and activities modelled by the ELH given in Figure C3 below.

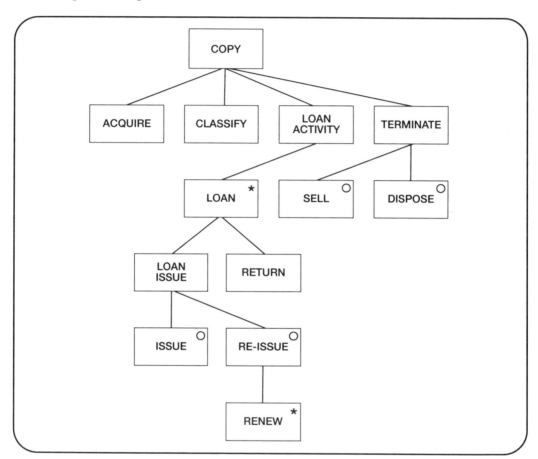

Figure C3

CHAPTER 10

Q1 Redraw the Logical Data Structure for Woodland Software Services using the Chen entity-relationship notation.

Q2 What are the advantages and disadvantages of using application packages to define system requirements?

Q3 Evaluate the current state of object oriented analysis.

Q4 Compare and contrast three different data modelling notations. Candidate standards might include:

> Bachman
> Martin (Information Engineering)
> Chen
> SSADM version 3
> SSADM version 4.

Index